STEAM Meets Story

STEAM Meets Story

Using Adolescent Fiction and Film to Spark Deeper Learning

Gloria D. Campbell-Whatley
Diane Rodriguez
Jugnu Agrawal

TEACHERS COLLEGE PRESS

TEACHERS COLLEGE | COLUMBIA UNIVERSITY

NEW YORK AND LONDON

Published by Teachers College Press®, 1234 Amsterdam Avenue, New York, NY 10027

Copyright © 2021 by Teachers College, Columbia University

Library of Congress Cataloging-in-Publication Data is available at loc.gov

ISBN 978-0-8077-6544-9 (paper)
ISBN 978-0-8077-6545-6 (hardcover)
ISBN 978-0-8077-7961-3 (ebook)

Printed on acid-free paper
Manufactured in the United States of America

Contents

Preface   ix

1. STEM and Literature: Exploring New Options   1
Gloria D. Campbell-Whatley and Diane Rodriguez

From Mainstreaming to Inclusion: The Journey Begins   2

The Birth of Multitiered Support Systems   2

Students With Disabilities in the Urban Environment   4

The Benefits of STEM and Literature   5

Summary and Other Tidbits   7

2. Culturally Responsive Problem-Based Learning (CRPBL)   13
Gloria D. Campbell-Whatley and Richard Reynolds

PBL Best Practices: Finding Solutions   14

Innovations in Instruction for PBL in STEM   14

Culturally Responsive Teaching   17

Culturally Responsive PBL (CRPBL)   18

Summary: Recommendations for Teachers   19

3. Selection of Reading Materials, Technology, and STEM Activities   23
Diane Rodriguez, Jugnu Agrawal, Justin Coles, and Gary Hoag

Benefits of Connecting Literature and STEM   23

Using Literature to Teach STEM in the Interdisciplinary Classroom: Selecting a Book   24

Using High-Leverage Practices and Strategies for Students With Disabilities   28

Supporting Learning in the Virtual Environment/Distance Learning   29

Lesson Planning   30

Summary   31

4. Science and Literature   35
Gloria Campbell-Whatley, Kim Reddig, and Deondra Gladney

Science and Literature for Adolescents   35

Careers   35

Teacher Application Tips   37

Summary of Lessons   38

Lesson 4.1. Science Vocabulary: The *Twilight* Saga   38

Lesson 4.2. Science on Cells: *The Immortal Life of Henrietta Lacks*   40

Lesson 4.3. Science and Clean Water: *The Boy Who Harnessed the Wind*   43

**Lesson 4.4. Science, Medicine, and Drugs: *Percy Jackson and the
    Lightning Thief***   45

Lesson 4.5. Gravity, Weight, and Mass: *Ender's Game*   48

Lesson 4.6. Respiratory System: *After Earth*   50

Lesson 4.7. Genetics: *Gattaca*   53

Lesson 4.8. Energy: *The Matrix*   54

Lesson 4.9. Desert Plants and Animals: *Holes*   57

Lesson 4.10. Assistive Technology: *Star Wars*   58

5. Mathematics and Literature   63
*Jugnu Agrawal, Diane Rodriguez, Gary Hoag, Ozalle Toms, Ann Jolly,
    Elizabeth Reyes, and Ashley Voggt*

Mathematics and Adolescents   63

Teacher Application Tips   63

Careers   64

Lesson Plans   64

Lesson 5.1. Statistics and Probability: *To Kill a Mockingbird*   65

Lesson 5.2. Basic Algebra and Statistics: *The Hunger Games*   70

Lesson 5.3. Business Plan: *Born a Crime*   72

Lesson 5.4. Geometry and Calculus: *Hidden Figures*   75

Lesson 5.5. Mysticism of the Pyramids: *The Giver*   78

Lesson 5.6. Weather Mathematics: *The Chronicles of Narnia*   79

Lesson 5.7. Geometry: *Flatland*   80

6. Engineering and Literature   85
*Diane Rodriguez, Gloria Campbell-Whatley, Kevin Otero, Hassan Payano,
    Maria Payano, Eileen Interiano, and Sharon Hunter*

Engineering for Adolescents   85

Careers   86

Tips for Teachers   86

Summary of Lessons   87

Lesson 6.1. Engineering Robotics: *The Maze Runner*   88

Lesson 6.2. Engineering Construction and Building: *Harry Potter*   90

Lesson 6.3. Drone Engineering: *Men in Black*   92

Lesson 6.4. Engineering Trains: *Snowpiercer*   94

Lesson 6.5. Satellite Engineering: *Star Trek*   97

Lesson 6.6. Engineering Cyborgs: *The Terminator*   99

Lesson 6.7. Software Engineering: *Jumanji*   101

Lesson 6.8. Aerospace Engineering: *Minority Report*   103

Lesson 6.9. Engineering and Space Travel: *The Martian*   106

Lesson 6.10. Biofuels: *Back to the Future*   107

7. Technology and Literature   **111**

Jugnu Agrawal, Gary Hoag, and Wen-Hsuan Chang

Technology-Related Careers   111

Teacher Application Tips   112

Summary of the Unit   112

Lesson 7.1. Identifying Credible Sources   112

Lesson 7.2. The Use of Citations and Presentations   114

Lesson 7.3. Making Use of Digital Media   116

Lesson 7.4. Using Technology to Update Writing Products   119

Lesson 7.5. Digital Citizenship and Technology Integration   121

Lesson 7.6. Webcams and Identifying Central Themes   123

Lesson 7.7. Digital Footprint   125

Lesson 7.8. Assistive Technology Innovations   127

Lesson 7.9. Digital Storytelling   129

Lesson 7.10. 3D Print   130

Index   **133**

About the Authors   **141**

To view the Common Core Standards that correspond with each of the lessons, as well as downloadable full-color handouts and a listing of related online resources, please visit the *STEAM Meets Story* page on www.tcpress.com and click on the Resources tab.

Preface

America's schools generally separate children using numerous labels. There are special and general education students, males and females, lower and higher socioeconomic levels, and various intellectual abilities. Many of these students could be successful in STEM fields because they have the talent, but they may lack the motivation or the wherewithal to know that they possess these skills. With this book, we hope to provide teachers with resources for creating lessons that let students know these skills can be acquired in an imaginative and innovative manner. In this text, we will use the arts, literature, and popular films to make this connection, thus using the cross-disciplinary techniques associated with STEAM (science, technology, engineering, arts, and math) to foster STEM success. Instruction in inclusive settings in general education classes needs to be adaptive and offer all students access and support. Linking literature with STEM and using movies and film as a conduit encourages opportunities to connect nonconcrete ideas to everyday academic experiences that students encounter.

This book strongly supports culturally and linguistically diverse (CLD) adolescents. Culturally relevant literature is infused into the disciplines as well as the lessons, along with standards-based instruction. The use of vocabulary strategies, concrete objects, and demonstrations are emphasized as these constructs and concepts help students learn. The book encourages academic equality and achievement, student participation, active learning, research to practice, the alignment of special and general education instructional goals and objectives, and inclusive practices that are especially important in urban settings.

Many general education teacher education programs provide limited options in instructional competencies to differentiate instruction, and this book can indeed assist them. *STEAM Meets Story: Using Adolescent Fiction and Film to Spark Deeper Learning* will engage both general and special education teachers in effective practical instructional Applications. They employ critical thinking through problem solving and problem-based learning techniques in STEM for adolescent students with ability differences in urban settings. This guide is grounded in current literature selected from the books most read by adolescent students. There are a number of STEM and reasoning skills that are explored, including *explicit instruction, inquiry-based learning,* and *problem-based learning,* and supported while developing the skills of analyzing and logical thinking.

The book helps guide teachers in learning techniques to stimulate interest in STEM careers for urban adolescents. Efforts to increase interest in STEM majors and careers begin at the middle school level. Many CLD adolescents in urban settings may not have the opportunity to be exposed to STEM careers and may not be encouraged to take precollege courses. *STEAM Meets Story: Using Adolescent Fiction and Film to Spark Deeper Learning* uses the Common Core State Standards (CCSS) because they have group standards that are used by many states and countries; can be easily matched to other state standards; and are interrelated and interwoven to pair science, mathematics, technology, and engineering with writing, reading, and speaking, as well as other subjects.

This book focuses on the STEM standards for adolescents in grades 6–12. The book is purposely written to be easy to read, practical, authentic, and not overly theoretical, and it includes strategies for the virtual environment. The chapters are written so that the reader can "cut to the chase" and readily get the gist of the ideas. In each familiar and popular story, saga, or series, English/language, science, mathematics, engineering, and technical subjects are combined to produce ideas and strategies that engage adolescents. Accommodations and strategies are embedded for diverse populations (e.g., students with differing abilities, traditionally underserved, and marginalized, transnational, and international students). In short, the book is structured to help teachers assist adolescents with skills to learn how to problem solve issues that they are likely to encounter. Chapters 1 through 3 provide the benefits of combining STEM with literature and the arts to foster STEM learning, and Chapters 4 through 7 contain lessons related to STEM. Separate chapters address science, technology, engineering, and math. Art is integrated throughout.

ACKNOWLEDGMENTS

We offer our genuine thanks and gratitude to all the writers who contributed to this book. Their collective wisdom demonstrates the power of cooperation and collaboration. This work is not a collection of independent contributions, but rather a gathering of individuals committed to a common belief in the possible. Collectively, they added immensely to the knowledge and expertise that is reflected in this book. These authors include colleagues and graduate students who are practitioners in the field of urban education and/or special education:

- Dr. Sharon M. Hunter, associate professor, North Carolina Agricultural and Technical College

- Dr. Ozalle Toms, assistant vice chancellor at the University of Wisconsin, Whitewater
- Dr. Richard Reynolds, director of student wellness and innovative programs at Warrensville Heights City Schools
- Dr. Justin Coles, assistant professor, curriculum and teaching, Fordham University
- Gary Hoag, president, Skillsmith, LLC
- Doctoral students practicing special education at the University of North Carolina at Charlotte: Wen-Hsuan Chang, Deondra Gladney, Ann Jolly, Kim Redding, Elizabeth Reyes, and Ashley Voggt
- Doctoral students from Fordham University: Maria Sol Anyosa, Eileen Interiano, Kevin Otero, Maria Payano, and Hassan Payano

Gloria Whatley-Campbell, EdD (Illustrator)
Diane Rodríguez, PhD
Jugnu Agrawal, PhD

STEAM Meets Story

CHAPTER 1

STEM and Literature

Exploring New Options

Gloria D. Campbell-Whatley and Diane Rodriguez

Educating adolescents with mild disabilities in inclusive environments can be quite challenging. Approximately 85% of students with disabilities have difficulties learning and struggle in science, technology, engineering, and mathematics (Camara & Quenemoen, 2012; U.S. Department of Education, 2004). Most students with mild disabilities have a learning disability, an attention deficit disorder, a speech/language disorder, or some other *academic-related disability* (Cawley et al., 2001; Henley et al., 2010). Reading and math deficits are by far the most common characteristics of academic-related disabilities, which are evident in 90% of these students (Kavale & Forness, 2003). These deficits complicate the understanding of informational text in science and technology, such as numerical reasoning, calculation, and other STEM-related skills, and pose major problems for students with a disability; consequently, they perform at lower levels than the norm (Cawley et al., 2001). Teachers may implement a flexible curriculum to help all students, including those at the margins (Rose et al., 2005). Meaningful practice with feedback usually improves their performance (Fuchs et al., 2003), and linking literature with STEM permits opportunities and advantages to connect abstract ideas to realistic experiences (Burns & Sheffield, 2004; National Council of Teachers of Mathematics [NCTM], 2000).

The educational challenge of the 21st century is to provide an equal opportunity for school success to all children. One third of students with disabilities are English learners (ELs) or are from diverse backgrounds. According to the National Center for Education Statistics (2020), one of every three students enrolled in public school is a student of a racial or ethnic minority background. One in five children under 18 lives in poverty, and one in seven children between the ages of 5 and 17 speaks a language other than English at home. Over 10 years, the number of ELs in America's classrooms has increased by 65%, causing greater demand for English as a second language (ESL) programs and services (Samson & Collins, 2012). Twelve percent of Black children in the United States were in classes for the disabled while only 8.5% of White children received those services (Barshay, 2019).

There is an acute need to integrate culturally relevant literature into disciplines, as well as accommodate language differences and cultural nuances, especially in the core curriculum and standards-based instruction (Santos et al., 2012). Further, communication among students, especially ELs, is critical for students to share their understanding of STEM principles and skills. Boyd-Batstone (2013) and Lee and colleagues (2013) emphasize the importance of using concrete objects and demonstration during the instruction of STEM practices in the classroom as these constructs help ELs learn STEM language.

This chapter introduces the combination of STEM and literature for culturally and linguistically diverse (CLD) adolescents in urban inclusive environments and will engage teachers in effective practical applications that employ critical thinking and problem-based learning techniques for STEM education for adolescents with mild disabilities. To date, a vast majority of adolescents with disabilities continue to be underserved in the implementation of STEM, in part because most of these classrooms continue to use traditional methods for teaching STEM. This book offers a viable, interesting, and stimulating way to present ideas, lessons, and examples based on famous adolescent books while using popular films as a conduit for understanding STEM concepts. This cross-disciplinary learning is often associated with the concept of STEAM (science, technology, engineering, arts, and math).

FROM MAINSTREAMING TO INCLUSION: THE JOURNEY BEGINS

The number of students with special needs receiving services in regular education settings has significantly increased over the past 3 decades. In the 2004 reauthorization of the Individuals with Disabilities Education Act (IDEA), Congress mandated preferred placement of students with disabilities in "an inclusive environment as an active participant" rather than just "being physically placed" through regular mainstreaming processes. Odom et al. (2012) stated that inclusion be implemented to help create an environment of involvement and connection in systems within schools and communities rather than be regarded as a physical placement. However, federal legislation did not provide a clear definition of inclusion. Consequently, when students with disabilities are placed hastily into general education settings, instructional experiences are the same as if they were mainstreamed.

Models that have students leave the general education setting to a separate room, such as the resource room, and other pull-out models in the continuum of special services, were challenged. Schools began attempting to implement the inclusion model without adequate professional training, a clear understanding of the concept of inclusion, and appropriate approaches and methods. Therefore, students did not receive the services they needed. A qualitative study by Vaughn and Shumm (1995) revealed that many teachers did not feel they had the appropriate skills, approaches, and training to effectively instruct students with disabilities and that instructional modifications were not feasible in the general education setting. Studies such as this one support the need for increased teacher experiences related to inclusion.

Inclusion became an emotionally charged topic, and a number of professional opinions regarding inclusion surfaced. Models of service, such as inclusion versus full inclusion, give general education and special education teachers different responsibilities (Blackman, 1992; Vaughn & Schumm, 1995). Presently there are few studies on the effects of inclusion and full inclusion programming, but there were studies that suggested students with disabilities were not receiving appropriate instruction in general education settings (Fuchs et al., 1993).

Moving Forward: Shooting Long and Going Far

The logistical difficulties of including diverse students in general education classrooms today are still apparent throughout the country (Witzel & Clarke, 2015). Successful implementation of inclusive procedures largely depends on teacher implementation (Witzel & Clarke, 2015). That is, the content, its presentation, and the collaboration between special and general educators is encouraged. Research on teachers working in inclusive settings has demonstrated that they still have serious reservations about including students with disabilities in their classrooms (Ring & Travers, 2005). Teachers vary significantly in their ability and willingness to make adaptations (McLeskey & Waldon, 2002). In their study, McLeskey and Waldon (2002) revealed that while some teachers stressed the importance of curricular and instructional adaptations, others reported ongoing difficulties in making all of the necessary adaptations in order to meet the needs of students with disabilities. Moreover, there seems to be a growing consensus of a slow progression toward educating students with disabilities in less restrictive settings. Therefore, it seems that teachers still need to provide more quality instruction for these students in the general education setting (Prasse et al., 2012; Witzel & Clarke, 2015).

Currently, 68.2% of students with disabilities are educated 80% of the day in the general education classroom (U.S. Department of Education, 2015). Creating more effective strategies for students with disabilities is paramount (Cosier et al., 2013; Kirby, 2017; Tremblay, 2013). Studies by Cosier and colleagues (2013), Shifrer (2013), and Tremblay (2013) show that students in inclusive settings have more educational gains in grade school and positive academic experiences in secondary education settings. Students who have the same opportunities as their peers in the general education setting have access to college preparatory courses, resulting in a better preparation for secondary schools. The learning environment and access to instructional resources have an impact on student performance (Lackaye & Margalit, 2006).

THE BIRTH OF MULTITIERED SUPPORT SYSTEMS

As larger numbers of CLD students enter our schools, the changing demographics have a major impact on our educational system (Ortiz & Yates, 2008). General and special educators face significant challenges to educate CLD students with disabilities. Understanding how cultural differences can affect the teaching/learning process is vital in order to provide responsive instruction (Kwon et al., 2017; Rodriguez, 2009). Professional development focused on teachers gaining an in-depth understanding of the influence of culture and language on students' academic performance is essential. Collaboration of educational professionals is important to inclusive settings in order to provide equitable opportunities to CLD students with disabilities (Roache et al., 2003). The lack of understanding may be why many CLD students, especially those in urban settings, were disproportionately placed in special education classes.

The National Association of School Psychologists (2013) defines disproportionality as a group's representation in a specific category that surpasses and differs significantly from the representation of others in that category. In other words, special education disproportionality has been referred to as "the extent to which membership in a given group affects the probability of being placed in a specific disability category" (Oswald et al., 1999, p. 198). Latinxs and African Americans have historically been disproportionately identified as in need of special education services and placed in more restrictive special educational settings.

Some special education law has addressed these concerns, such as the Education for All Handicapped Children Act (EHA), which supported an intelligence test as the basic assessment instrument used to determine eligibility for special education. However, this tool became problematic (Deno, 1987; Heron & Heward, 1982; Karier, 1972; *Larry P. v. Riles*, 1976; *Lora v. The Board of Education*, 1977).

These types of tests have an extensive history. Alfred Binet created the first intelligence test. Campbell-Whatley and Comer (2000) pointed out that the test was originally designed to recognize students who might benefit from supplementary academic assistance. In the United States, Goodard (1920) and Terman (1916) modified the intelligence test to devise a differentiated curriculum for the theoretical intellectual elite. They believed that measured intelligence reflected genetic inheritance (Karier, 1972). Additionally, it believed that the test favored socioeconomically privileged students, and many versions of the test were heavily based on verbal items (Karier, 1972). In addition, many timed tests were administered by evaluators who were not familiar with the student's vernacular or language and had an inaccurate view of what the student knew.

There are other challenges with intelligence tests. The use of intelligence testing to place special education students from culturally and linguistically diverse (CLD) backgrounds was the major cause of disproportionately placed students (Artiles et al., 2010; Blanchett, 2006; Harry & Klinger, 2006; Moody, 2016; U.S. Commission on Civil Rights, 2009). These placements were often the result of cultural differences rather than true deficiencies in student learning and achievement (Ortiz & Yates, 1983). Such culture-based differences may be misinterpreted as a type of disability category, resulting in educators disproportionately recommending their CLD students for special education services (Oswald et al., 1999; Turnbull et al., 2009).

CLD and Response to Intervention and Multitiered Support Systems: A Different Alphabet Soup

Response to intervention (RTI) and multitiered support systems (MTSS), proactive approaches that identify and support students with learning and behavior needs through the use of research-based intervention, were created to reduce bias in determining eligibility for special education. The change was prompted by IDEA 2004 because of the overrepresentation of CLD students in special education, but it also was designed to help all students through the use of research-based interventions as soon as their at-risk status was determined. The type of instruction is also significant for the effectiveness of RTI/MTSS.

Studies have assessed both individual and small group instruction, and there are some studies that show the effectiveness of RTI/MTSS for students (Witzel & Clarke, 2015). Prasse and colleagues (2012) found that 56% of general and special educators are working together to implement RTI/MTSS systems that include curriculum differentiation. Although there is a 49% increase in the use of RTI/MTSS and a reduction in special education referrals (Frasse, et al., 2012; Global Scholar, 2011), problems remain for African American and Latinx students. African Americans remain overrepresented in special education and are about 1.5 times more likely to receive special education services than same-age students of all other racial/ethnic groups combined (U.S. Department of Education, 2010).

Artiles (2014) argues that the current educational research methodology employed to study the efficacy of RTI/MTSS and other special education programs is incomplete because it fails to fully embrace a more comprehensive, multifaceted definition of culture, and, in so doing, falls short of what it could provide for students. Consequently, the model lacks cultural competency because it does not connect differences in race, ethnicity, socioeconomic status, gender, and ability. Other researchers support this premise. For example, McMahon and colleagues (2016) performed a study of a midsized urban school district whose classroom best practices focused on the following: (a) the value of student participation, (b) active learning, (c) research to practice, (d) the alignment of special and general education instructional goals and objectives, and (e) inclusion in and out of the classroom. These researchers found that inclusion was particularly important in urban areas, especially for students from CLD backgrounds who contend not only with a disability but with poverty and discrimination. Like Artiles, they found that it is important to consider other variables to ensure student success. Teacher preparation is significant to student success

Many general education teacher education programs are limited in preparation and instructional competencies to differentiate instruction, perform necessary assessment, and appropriately monitor progress data, and these variables are directly related to student knowledge (Prasse et al., 2012). Prasse and colleagues (2012) surveyed general education teachers on their ability to instruct and assess students with mild disabilities. Less than 40% of those teachers believed that they could not achieve academic benchmarks and 76% said they needed support in instructing and developing interventions for students with mild disabilities. The Zigmond and colleagues' (2011) study, which took place in low-achieving schools in urban settings, demonstrated a direct relation between teacher instructional knowledge and student performance. Clemens and colleagues (2012) conducted a study in which MTSS was implemented across a district. Student outcomes did not improve when general education teachers declined to implement the MTSS programs that were fashioned for them. It is imperative that students in urban settings receive differentiated, scaffolded instruction and that teachers become able to provide that form of instruction in general education classrooms (Mason & Hedin, 2011; Prasse et al., 2012).

Due to the cultural shift among students in the United States, teachers and school leaders are expanding their knowledge and skills to support ELs' language and literacy development and to reliably identify and effectively educate ELs who have a language or reading disability. IDEA has always included specific guidelines for determining eligibility for students with disabilities; therefore, it is suggested that stakeholders in this era be knowledgeable about the law, which states, "In making a determination of eligibility under paragraph (4)(A), a child shall not be determined to be a child with a disability if the determinant factor for such determination is

(A) lack of appropriate instruction in reading, including in the essential components of reading instruction (as defined in section 6368 (3) of this title);
(B) lack of instruction in math; or
(C) limited English proficiency. (20 U.S.C. 1414(b)(5), section 34 CFR 300.306(b))

STUDENTS WITH DISABILITIES IN THE URBAN ENVIRONMENT

Sixty-five million students are served in urban centers. A school district is considered as urban when it has 70% of its schools in the city. These schools generally have relatively high poverty (i.e., free and reduced lunch data, a large proportion of students of color and ELs) (Garza, 2009). Given the racial disparities in achievement and economic levels, it has become necessary that instructional practices pay attention to culturally relevant pedagogy (CRP) (Ladson-Billings, 1995). CRP encourages teachers to link home, school, and community (Brown-Jeffy & Cooper, 2011). The challenge for teachers will be integrating these variables into instruction. At the core of this challenge is navigating change and finding ways to raise the level of learning in the classroom. The National Council of Teachers of English (2016) published a position statement addressing these issues. In order to support urban adolescents, they believe the following:

- Teachers respect all learners and themselves as individuals with culturally defined identities.
- Students bring their own knowledge into their learning communities, and, recognizing this, teachers can incorporate this knowledge and experience into classroom practice.
- Students have a right to a variety of educational experiences that help them make informed decisions about their role and participation in language, literacy, and life.
- Teachers need to model culturally responsive and socially responsible practices for students.

Feistritzer (2011) emphasizes that culture does matter and is a part of the learning environment. To be culturally competent means that a teacher can successfully instruct someone from a culture other than their own. Cultural competence is a readiness to develop interpersonal relationships, learn the student culture, and design applicable experiences for the student (Brown-Jeffy & Cooper, 2011). The next challenge is adjusting instructional practices so that culture has relevancy across all content areas.

Ladson-Billings (1995) outlines foundations for CRP. Included is the idea formation of self and others, knowing the social relations of students, and the foundations of knowledge for varied cultures. It is encouraged that teachers believe that all students are capable of academic success and that there is a means to connect with students. If possible, teachers can invest in positive educational outcomes for all learners in order to provide a comprehensive inclusive model in which students have opportunities to succeed and learn about the subject matter. The following strategies can be applied to CLD students and STEM education (Haager &Vaughn, 2013; Ladson-Billings, 1995):

- Students can implement small group instruction with guided practice within integrated STEM lessons.
- STEM instruction can be differentiated for skill level and for culturally responsive instruction.
- Class assignments should have multiple ways students can respond and demonstrate knowledge of STEM objectives.

THE BENEFITS OF STEM AND LITERATURE

The STEM coalition was formed in response to former president Barack Obama's call to action to improve educational outcomes, encourage students to seek STEM careers, and raise student awareness of the importance of science and technology to our nation's future. A national lab network of STEM professionals began working in partnership with teachers and schools to contribute knowledge and skills to improve secondary school facilities to support STEM learning in a hands-on environment. Linking STEM ideals with literature is a nontraditional approach to connect learning with technical research and pedagogical approaches to teaching, and studies have linked mathematics, science, and literature to an effective teaching and learning strategy for STEM (McKinney et al., 2017; Tucker et al., 2014).

In light of the success linking mathematics, science, and literature for instructional purposes, a new question emerges: How do we do this for adolescents with mild disabilities? Many of these students will attend college and have the ability to partake in STEM careers if they are motivated and

the right strategies are used to instruct them in general and special education settings. Due to MTSS, adolescents with mild disabilities spend about 80% of their time in general education classrooms. They struggle with subjects related to STEM, not only with reading, but technical terms and challenging vocabulary words. Usually, in STEM-related subjects, only half of adolescents with mild disabilities reach average to above average proficiency levels (Thurlow et al., 2010). STEM texts have four challenging characteristics for adolescents with disabilities (Sáenz & Fuchs (2002):

- *STEM texts have a more complex structure as compared to a reading text.* Simply locating the main idea of the text is challenging when multiple organization patterns are present within a single paragraph.
- *STEM texts are conceptually dense; to be exact, they are loaded with concepts and logical/causal relationships.* These concepts become more and more abstract as adolescents move to higher grade levels, intensifying difficulty in interpretation.
- *The vocabulary is challenging.* Vocabulary knowledge deters students from making connections to novel and technical vocabulary and to infer meaning when reading the text.
- *Organization patterns do not help make connections to prior knowledge.* Due to the complexity of organizational patterns, some students cannot connect what was learned in the past to what is currently being taught.

Overall, the reading demands of STEM texts place these students at a great disadvantage. Yet use of validated strategies and scaffolded instruction can help many adolescents with disabilities. Root-Berstein (2012) reflected on applicable strategies for adolescent students in urban settings with mild disabilities. Newer technologies, including functional magnetic resonance imaging (fMRI), allow real-time analysis of brain activity on the different modalities of learning: visual, auditory, tactile, and kinesthetic. If an individual is taught a concept through a series of images that are accompanied with auditory descriptions, students are more likely to recall the concept. If the concept is illustrated visually, the learner has more access points for that information. When an individual is taught a single concept, the brain creates neural pathways connecting that concept to their experience. The more access points or neural pathways established, the greater the chance of retention and recall.

There has been discussion, debate, and investigation on how the brain organizes language in English learners and whether two different languages are localized in the same or in different areas of the brain (Dehaene-Lambertz et al., 2008; Petitto, 2009). The dialogue has attempted to answer questions such as the following: How do the functions of the English learner's brain compare to the monolingual brain? Given the recent studies in brain resonance imaging, educators are taking an interest in learning more about important implications for "understanding core properties of all human language and reading, challeng[ing] assumptions about language and reading as being tied to sound, and provid[ing] novel insight into a remarkable biological equivalence in signed and spoken languages" (Petitto et al., 2016, p. 1). Integrating the literature into core content areas enables urban CLD adolescents with disabilities to explore a single concept from different vantage points, and it also utilizes all the different modalities of learning, and both lead to the formation of more neural pathways. Brain research supports the multidimensional methods to use literature to teach STEM.

Motivation is also a significant variable relating to student success. A literature review by Aeschlimann (2016) revealed four aspects that can increase motivation in urban adolescents with disabilities in STEM classes:

- *Provide information about STEM career opportunities.* Since students with mild disabilities, especially those in urban settings, benefit from encouragement to pursue STEM fields, teachers may inform students that employment in STEM occupations is attainable for all. Adolescents are empowered when they are provided career role models, speakers, and mentors to improve confidence.
- *Ensure comprehensible STEM teaching.* The material for adolescents with disabilities can be scaffolded with plentiful examples accompanied by sufficient time to absorb the material. For better comprehension, the format of presentations can be varied to allow for multiple means of expression, engagement, and presentation.
- *Provide individual support to students.* A good student–teacher relationship is essential, especially for students with disabilities in STEM classes. Adolescents with disabilities, just as any student, need to be in supportive environments.
- *Connect subject-specific matters with everyday experience.* Adolescents with disabilities from urban environments benefit from examples and illustrations that come from activities that are familiar to real-life day-to-day contexts.

Agrawal and Morin (2016) emphasized evidence-based practice for urban adolescents with disabilities that are applicable to STEM concepts. The concrete representational abstract (CRA) framework provides teachers with specific procedures to guide students through STEM concepts: First, hands-on components encourage the student to use manipulatives. Next, the use of visual representation bridges the gap between tactile/kinesthetic and written concepts. The student then demonstrates the ability to comprehend through expression or presentation of the skills.

McKinney and colleagues (2017) outlined several advantages for linking STEM and literature that can be applied to students with disabilities in urban inclusive settings: (a) developing STEM concepts and vocabulary, (b) promoting problem-solving and reasoning skills, and (c) stimulating interest in stem careers.

Developing STEM concepts and vocabulary skills using literature selections can serve to increase students' understanding of vocabulary words. As mentioned earlier in this chapter, students with disabilities have difficulty with STEM concepts (Camara & Quenemoen, 2012; U.S. Department of Education, 2004) and combining STEM and literature is one way to stimulate interest. For example, *Twilight* starts a series of four vampire-themed fantasy romance novels that tell about the life of Bella, a teenage girl who falls in love with a vampire named Edward. As illustrated in Chapter 4 of this resource, students can view the YouTube clip from the first *Twilight* movie, "Twilight Biology Class Scene Edward's Golden Eyes," and list and define the science vocabulary words. Additionally, they can discuss biology-related concepts such as cold-blooded and warm-blooded animals, bats and their habitat, and animals that drink blood. Interrelated skills such as vocabulary use in *writing* and *literacy* strengthen the use of (a) domain-specific vocabulary; (b) specialized reference materials (e.g., dictionaries, glossaries, thesauruses); and (c) figurative language, word relationships, nuances in word meanings. Different media (e.g., print or digital text, video, multimedia) can be used to present and emphasize concepts.

Stem combined with literature can promote problem-solving and reasoning skills. *Inquiry-based learning* and *problem-based learning* are supported while developing the skills of analysis and logical thinking (Archer & Hughes, 2010; Goeke, 2008). Inquiry-based learning is well suited to the scientific process because it encourages investigation by posing problem-solving questions or scenarios rather than presenting established facts. Properly posed questions can serve as a formative measure to assess a student's knowledge and understanding of a STEM activity. Inquiry-based learning works hand-in-hand with problem-based learning (PBL) methods used in STEM, which involve research exploration that develops practical, logical, and relevant solutions.

For example, the book *Holes* (Sachar, 1998) emphasizes several problem-solving and reasoning skills. For example, Stanley Yates has been unjustly sent to a boys' detention center, Camp Green Lake, in a desert area, where the warden makes the boys "build character" by spending all day, every day, digging holes 5 feet wide and 5 feet deep. It doesn't take long for Stanley to realize there's more than character improvement going on at Camp Green Lake (problem solving). The boys are digging holes because the warden is looking for treasure. The movie reveals the medicinal uses of the onion (e.g., hair growth) that grow wild in that area. Stanley's dad has spent his life attempting to find a cure for smelly shoes (caused by *Brevibacterium*), conducting countless odor-eliminating experiments. He finally cures smelly shoes with Sploosh, a combination of onions and peaches (problem-based learning). Clips of the movie show Stanley and his friend Zero covered with the deadly venomous, yellow-spotted lizards, but they are not bitten due to their consumption of onions because the onion acts as repellant to the spotted reptiles (inquiry based). Inquiry-based learning is further supported when other plants that act as a repellant are discussed (i.e., citronella).

McKinney and colleagues (2017) point out that stimulating interest in STEM careers is significant. Labor statistics from the United States predict the creation of 3 million new jobs in STEM by 2020 (U.S. Bureau of Labor Statistics, 2010), and efforts to increase student interest in STEM careers have been on the rise. These efforts typically begin at the middle school level. There are a number of studies that mention ways to increase interest for underrepresented groups in STEM careers (Blanchard et al., 2012; Yoon & Strobel, 2017). Many CLD adolescents in urban settings may not have the opportunity to be exposed to STEM careers; as mentioned before, they would not, in general, be encouraged to take precollege courses. Yoon and Strobel (2017) found in a study on adolescent trends related to STEM courses in high school that Latinx and African American adolescents were less likely to take mathematics and science courses compared to White students. Nevertheless, if they were encouraged and took the courses, the CLD adolescents tended to be just as likely to seek STEM degrees as White students.

SUMMARY AND OTHER TIDBITS

This chapter introduced and explored inclusion and the benefits of integrating STEM and literature for adolescents with disabilities in urban environments. History demonstrates that the special education continuum of services moved from mainstreaming to inclusion using RTI/ MTSS strategies. At the same time, changes in eligibility procedure methods progressed from the IQ test to tiered supports, although disproportionate placement is still problematic for African American and Latinx populations. Finally, the many strengths of integrating STEM and literature were outlined. The chapters that follow will

- offer literature selections that can be used as catalysts to launch student interactions, discussions, and investigations that emphasize STEM processes, such as problem-solving and reasoning skills in inclusive environments;
- use research-based teaching strategies that are essential for the urban learners and those that are at risk or have mild disabilities;
- provide handouts for engaging, hands-on STEM activities; and
- present ideas for virtual environments.

Subsequent chapters will provide various strategies that lend themselves to integrating literature and STEM such as explicit instruction, inquiry-based learning, and PBL for adolescents with mild disabilities (Archer & Hughes, 2010; Goeke, 2008). These methods will assist adolescents in the application of real-world skills in a meaningful manner while examining problem-solving approaches and developing practical, logical, and relevant solutions to STEM-based problems.

REFERENCES

Aeschlimann, B., Herzog, W., & Makarova, E. (2016). How to foster students' motivation in mathematics and science classes and promote students' STEM career choice. A study in Swiss high schools. *International Journal of Educational Research*, 79, 31–41.

Agrawal, J., & Morin, L. L. (2016). Evidence-based practices: Applications of concrete representational abstract framework across math concepts for students with mathematics disabilities. *Learning Disabilities Research & Practice, 31*(1), 34–44.

Archer, A. L., & Hughes, C. (2010). *Explicit instruction: Effective and efficient teaching (what works for special-needs learners)*. Guildford.

Artiles, A., Harry, B. Reschly, D. & Chinn, P. (2010). Overidentification of students of color in special education: A critical overview. *Multicultural Perspectives, 4*(1), 3–10. https://doi.org/10.1207/S15327892MCP0401_2

Artiles, A. J. (2014). Beyond responsiveness to identity badges: Future research on culture in disability and implications for Response to Intervention. *Educational Review, 67*(1), 1–22. https://doi.org/10.1080/00131911.2014.934322

Blackman, H. P. (1992). Surmounting the disability of isolation. *The School Administrator, 49*, 28–29.

Blanchard, M. R., Albert, J. L., Alsbury, T. L., & Williams, B. (2012). *NSF ITEST annual project outcomes report: Innovative technology experiences for students and teachers. STEM teams: promoting science, technology, engineering, and mathematics (STEM) career interest, skills, and knowledge through strategic teaming*. National Science Foundation. https://www.researchgate.net/publication/259066990_The_Development_of_the_STEM_Career_Interest_Survey_STEM-CIS

Blanchett, W. (2006). Disproportionate representation of African American students in special education: Acknowledging the role of white privilege and racism. *Educational Researcher, 35*(6), 24–28.

Boyd-Batstone, P. (2013). *Helping English language learners meet the Common Core: Assessment and instructional strategies K–12*. Routledge.

Brown-Jeffy, S., & Cooper, J. (2011). Toward a conceptual framework of culturally relevant pedagogy: An overview of the conceptual and theoretical literature. *Teacher Education Quarterly, 38*(1), 65–84.

Burns, M., & Sheffield, S. (2004). *Math and literature*. Math Solutions.

Camara, W., & Quenemoen, R. (2012, January 5). *Defining and measuring college and career readiness and informing the development of performance level descriptors (PLDs)*. National Center on Educational Outcomes.

Campbell-Whatley, G. D., & Comer, J. (2000). Ethics, power, and privilege: Self-concept and African-American student achievement. *Teacher Education and Special Education, 23*, 19–31.

Cawley, J., Parmar, R., Foley, T. E., Salmon, S., & Roy, S. (2001). Arithmetic performance of students: Implications for standards and programming. *Exceptional Children, 67*(3), 311–328. https://doi.org/10.1177/001440290106700302

Clemens, N., Shaprio, E., Hilt-Panahon, A., & Gischlar, K. (2012). Student achievement outcomes. In E. S. Shaprio, N. Zigmond, T. Wallace, & D. Marston (Eds.), *Models for implementing response to intervention: Tools, outcomes, and implications* (pp. 77–98). Guilford.

Cosier, M., Causton-Theoharis, J., & Theoharis, G. (2013). Does access matter? Time in general education and achievement for students with disabilities. *Remedial and Special Education, 34*(6), 323–332. https://doi.org/10.1177/0741932513485448.

Dehaene-Lambertz, G. Hertz-Pannier, L., Dubois, J., & Dehaene, S. (2008). How does brain organization promote language acquisition in humans? *European Review, 16*(4), 399–411.

Deno, S. L. (1987). Curriculum-based measurement. *Teaching Exceptional Children, 20*, 41–42.

Feistritzer, C. E. (2011). *Profile of teachers in the U.S. 2011*. National Center for Education Information.

Fuchs, D., Fuchs, L. S., & Fernstrom, P. (1993). A conservative approach to special education reform: Mainstreaming through transenvironmental programming and curriculum-based measurement. *American Educational Research Journal, 30*, 149–177.

Fuchs, D., Mock, D., Morgan, P. L., & Young, C. L. (2003). Responsiveness-to-intervention: Definitions, evidence, and implications for the learning disabilities construct. *Learning Disabilities Research & Practice, 18*(3), 157–171.

Garza, R. (2009). Latino and white high school students' perceptions of caring behaviors: Are we culturally responsive to our students? *Urban Education, 44*(3), 297. https://doi.org/10.1177/0042085908318714

Global Scholar. (2011). *Response to Intervention Adoption Survey 2010*. Cision. https://www.prnewswire.com/news-releases/2011-response-to-intervention-report-by-globalscholar-nasdse-case-and-aasa-uncovers-latest-trends-in-rti-adoption-among-us-school-districts-128001008.html

Goeke, J. (2008). *Explicit instruction: A framework for meaningful direct teaching*. Pearson.

Goodard, H. H. (1920). *Human efficiency and levels of intelligence*. Princeton University Press.

Haager, D., & Vaughn, S. (2013). The common core state standards and reading: Interpretations and implications for elementary students with learning disabilities. *Learning Disabilities Research & Practice, 28*(1), 5–16.

Hang, Q., & Rabren, K. (2009). An examination of co-teaching: Perspectives and efficacy indicators. *Remedial and Special Education, 30*(5), 259–268. https://doi.org/10.1177/0741932508321018

Harry, B., & Klingner, J. (2006). *Why are so many minority students in special education? Understanding race and disability in schools.* Teachers College Press.

Henley, M., Ramsey, R. S., & Algozzine, R. F. (2010). *Common characteristics of students with mild disabilities.* Pearson.

Heron, T. E., & Heward, W. L. (1982). Ecological assessment implications for teachers of learning disabled students. *Learning Disabilities Quarterly, 11,* 224–232.

Individuals with Disabilities Education Act, 20 U.S.C. §1400 (2004).

Karier, C. J. (1972). Testing for order and control in the corporate liberal state. *Educational Theory, 22,* 154–180.

Kavale, K. A., & Forness, S. R. (2003). Learning disability as a discipline. In H. L. Swanson, K. R. Harris, & S. Graham (Eds.), *Handbook of learning disabilities* (pp. 76–93). Guilford.

Kirby, M. (2017). Implicit assumptions in special education policy: Promoting full inclusion for students with learning disabilities. *Child Youth Care Forum, 46,* 175–191.

Kwon, K., Hong, S., & Jeon, H.-J. (2017). Classroom readiness for successful inclusion: Teacher factors and preschool children's experience with attitudes toward peers with disabilities. *Journal of Research in Childhood Education, 31*(3), 360–378.

Lackaye, T. D., & Margalit, M. (2006). Comparisons of achievement, effort, and self-perceptions among students with learning disabilities and their peers from different achievement groups. *Journal of Learning Disabilities, 39*(5), 432–446. https://doi.org/10.1177/00222194060390050501

Ladson-Billings, G. (1995). But that's just good teaching! The case for culturally relevant pedagogy. *Theory into Practice, 34*(3), 159–165.

Larry P. v. Riles, 343 F. Supp. 169 (1976).

Lee, O., Quinn, H., & Valdés, G. (2013). Science and language for English language learners in relation to Next Generation Science Standards and with implications for Common Core State Standards for English language arts and mathematics. *Educational Researcher, 42*(4), 223–233. https://doi.org/10.3102/0013189X13480524

Lenz, B. K., Deshler, D. D., & Kissam, B. R. (2004). *Teaching content to all: Evidence-based inclusive practices in middle and secondary schools.* Pearson.

Lora v. The Board of Educ., 74 F.R.D. 565 (E.D.N.Y.1977).

Lucas, T. (2000). Facilitating the transition of secondary English language learners: Priorities for principals. *National Association of Secondary School Principals NASSP Bulletin, 84*(619), 1–16.

Mason, L. H., & Hedin, L. R. (2011). Reading science text: Challenges for students with learning disabilities and considerations for teachers. *Learning Disabilities Research & Practice, 26*(4), 214–222.

McKinney, S., Tomovic, C., Grant, M., & Hinton, K. (2017). Increasing STEM competence in urban, high poverty elementary school populations. *K-12 STEM Education, 3*(4), 267–281.

McLeskey, J., & Waldron, N. (2002). Professional development and inclusive schools: Reflections on effective practice. *The Teacher Educator, 37*(3),159–172. https://doi.org/10.1080/08878730209555291

McLeskey, J., Waldron, N. L., & Womhoff, S. A. (1990). Factors influencing the identification of Black and White students with learning disabilities. *Journal of Learning Disabilities, 23*(6), 362–367.

McMahon, S. D., Keys, C. B., Berardi, L., Crouch, R., & Coker, C. (2016). School inclusion: A multidimensional framework and links with outcomes among urban youth with disabilities. *Journal of Community Psychology, 44*(5), 656–673.

Moody, M. (2016). From under-diagnoses to over-representation: Black children, ADHD, and the school-to-prison pipeline. *Journal of African-American Studies, 20,* 152–163. https://doi.org/10.1177/2156869320916535

Morgan, P., Farkas, G., & Hillemeier, M. M. (2015). Minorities are disproportionately underrepresented in special education: Longitudinal evidence across five disability conditions. *Educational Researcher, 44*(5), 278–292.

National Association of School Psychologists. (2013). *Racial and ethnic disproportionality in education.* https://www.nasponline.org/resources-and-publications/resources-and-podcasts/diversity-and-social-justice/disproportionality

National Center for Education Statistics. (2020). *Racial/ethnic enrollment in public schools.* https://nces.ed.gov/programs/coe/indicator_cge.asp

National Council of Teachers of English. (2016, July 1). *Summary of NCTE policies to promote diversity and inclusion within the council.* http://www2.ncte.org/statement/nctediversity/

National Council of Teachers of Mathematics. (2000). *Standards for school mathematics.* Author.

Odom, S., Buysse, V., & Soukakou, E. (2012). Inclusion for young children with disabilities: A quarter century of research perspectives. *Journal of Early Intervention, 33*(4), 344–356.

Ortiz, A., & Yates, J. (2008) Enhancing scientifically-based research for culturally and linguistically diverse learners. *Multiple Voices for Ethnically Diverse Exceptional Learners, 11*(1), 13–23.

Oswald, D. P., Coutinho, M. J., Best, A. M., & Singh, N. N. (1999). Ethnic representation in special education: The influence of school related economic and demographic variables. *The Journal of Special Education, 32,* 194–206.

Petitto, L. A. (2009). New discoveries from the bilingual brain and mind across the life span: Implication for education. *International Journal of Mind, Brain, and Education, 3*(4), 185–197.

Petitto, L. A., Langdon, C., Stone, A., Andriola, D., Kartheiser, G., & Cochran, C. (2016). Visual sign phonology: Insights into human reading and language from a natural soundless phonology. *WIREs Cognitive Science, 7*(6), 366–381. https://doi.org/10.1002/wcs.1404

Prasse, D., Breunlin, R., Giroux, D., Hunt, J., Morrison, D., & Thier, K. (2012). Embedding multi-tiered system of supports/response to intervention into teacher preparation. *Learning Disabilities: A Contemporary Journal, 10*(2), 75–93.

Ring, E., & Travers, J. (2005). Barriers to inclusion. *European Journal of Special Needs Education, 20*(1), 41–56.

Roache, M., Shore, J., Gouleta, E., & Butkevich, E. (2003). An investigation of collaboration among school professionals in serving culturally and linguistically diverse students with exceptionalities. *Bilingual Research Journal, 27*(1), 117–136.

Rodriguez, D. (2009). Meeting the needs of English language learners with disabilities in urban settings. *Urban Education, 44*(4), 452–464.

Root-Bernstein, R. (2012). Arts foster scientific success. *Journal of Psychology of Science and Technology, 1*(2), 51–63.

Rose, D., Meyer, A., & Hitchcock, C. (2005). Introduction. In D. Rose, A. Meyer & C. Hitchcock (Eds.), *The universally designed classroom: Accessible curriculum and digital technologies* (pp. 1–12). Harvard Education Press.

Sachar, L. (1998). *Holes.* Farrar, Straus and Giroux.

Sáenz, L. M., & Fuchs, L. S. (2002). Examining the reading difficulty of secondary students with learning disabilities: Expository versus narrative text. *Remedial and Special Education, 23*(1), 31–41. https://doi.org/10.1177/074193250202300105

Samson, J. F., & Collins, B. A. (2012). *Preparing all teachers to meet the needs of English language learners: Applying research to policy and practice for teacher effectiveness.* Center for American Progress. https://files.eric.ed.gov/fulltext/ED535608.pdf

Santos, M., Darling-Hammond, L., & Cheuk, T. (2012). Teacher development to support English language learners in the context of common core state standards. *Understanding language: Language, literacy, and learning in the content areas.* http://ell.stanford.edu/sites/default/files/pdf/academic-papers/10-Santos%20LDH%20Teacher%20Development%20FINAL.pdf

Shifrer, D. (2013). Stigma of a label: Educational expectations for high school students labeled with learning disabilities. *Journal of Health and Social Behavior, 54*(4), 462–480. https://doi.org/10.1177/0022146513503346.

Strogilos, V., & Avramidis, E. (2016). Teaching experiences of students with special educational needs in co-taught and non-co-taught classes. *Journal of Research in Special Educational Needs, 16*(1), 24–33. http://dx.doi.org/10.1111/1471-3802.1205

Terman, M. (1916). *The measurement of intelligence.* Houghton Mifflin.

Thurlow, M., Rogers, C., & Christensen, L. (2010). *Science assessments for students with disabilities in school year 2006–2007: What we know about participation, performance, and accommodations* [Synthesis report 77]. National Center on Educational Outcomes.

Tremblay, P. (2013). Comparative outcomes of two instructional models for students with learning disabilities: Inclusion with co-teaching and solo-taught special education. *Journal of Research in Special Educational Needs, 13*(4), 251–258. https://doi.org/10.1111/j.1471-3802.2012.01270.x

Tucker, C., Boggan, M., & Harper, S. (2010). Using children's literature to teach measurement. *Reading Improvement, 47*(3), 154–161.

Turnbull, R., Huerta, M., & Stowe, N. (2009). *What every teacher should know about: The Individuals with Disabilities Education Act as amended in 2004* (2nd ed.). Pearson.

U.S. Bureau of Education for the Handicapped. (2007). *Progress toward a free appropriate public education; a report to congress on the implementation of Public Law 94-142: The Education for All Handicapped Children Act.* https://www2.ed.gov/policy/speced/leg/idea/history.html

U.S. Bureau of Labor Statistics. (2010). *Occupational outlook handbook, 2010–11 library edition.* https://fraser.stlouisfed.org/title/occupational-outlook-handbook-3964/occupational-outlook-handbook-2010-11-edition-583142/content/pdf/bls_2800_2010_libraryedition

U.S. Commission on Civil Rights. (2009). *Minorities in special education.* https://www.usccr.gov/pubs/docs/MinoritiesinSpecialEducation.pdf

U.S. Department of Education. (2004). *The condition of education.* National Center for Educational Statistics. https://nces.ed.gov/pubs2004/2004077.pdf

U.S. Department of Education. (2010). *The 29th annual report to Congress on the implementation of the Individuals with Disabilities Education Act, 2007: Vol. 2.* https://

www2.ed.gov/about/reports/annual/osep/2007/parts
-b-c/29th-vol-2.pdf

Vaughn, S., & Shumn, J. S. (1995). Responsible inclusion for students with learning disabilities. *Journal of Learning Disabilities, 28*(5), 264–270.

Witzel, B., & Clarke, B. (2015). Focus on inclusive education: Benefits of using a multi-tiered system of supports to improve inclusive practices. *Childhood Education, 91*(3), 215–219. https://doi.org/10.1080/0 0094056.2015.1047315

Yoon, S. Y., & Strobel, J. (2017). Trends in Texas high school student enrollment in mathematics, science, and CTE-STEM courses. *International Journal of STEM Education, 4*(1), 9.

Zigmond, N., Kloo, A., & Stanfa, K. (2011). Celebrating achievement gains and cultural shifts. In E. Shapiro, N. Zigmond, T. Wallace, & D. Marston (Eds.), *Models for the implementation of response to intervention: Tools, outcomes, and implications* (pp. 171–198). Guilford.

CHAPTER 2

Culturally Responsive Problem-Based Learning (CRPBL)

Gloria D. Campbell-Whatley and Richard Reynolds

The incorporation of culturally responsive problem-based learning allows practitioners to develop lessons that do not alienate students' cultural experiences. Focusing on student deficits and disabilities permits and encourages the perpetuation of the achievement gap. This chapter defines problem-based learning and its impact on diverse students, on culturally responsive teaching, and the coupling of the two strategies. Factors such as race, school inequities, and best practices for instruction for diverse students and how the use of these approaches culturally impacts learning will be addressed.

There are six core characteristics of *problem-based learning* (PBL) (Barrows, 1996; Dochy et al., 2003):

1. Student-centered learning
2. Small group instruction
3. Facilitator-led workgroups
4. Real-life problems
5. Self-directed structured problem solving framework
6. The development of problem-solving skills

In group format, PBL structuring stimulates students to use prior information to problem solve. The facilitator (i.e., teacher, educator) assists and leads the group and helps them recognize gaps in the information provided so they can develop a solution. Students play an active part in learning by devising or answering questions to guide learning activities. Cognitive engagement like this allows for the initiation and stimulation of prior knowledge as well as encouraging critical analysis, thereby promoting an increased understanding of the problem (Loyens et al., 2015).

Cognitive engagement is the quality of higher order thinking strategies, as demonstrated by the hierarchy of Bloom's taxonomy (Shabatura, 2020). Students can engage in order to become more self-directed in the learning environment and intensify the motivation to participate in the focused tasks required by PBL (Linnenbrink, 2007).

Researchers agree that best practices actively engage students through instructional activities that encourage interactions with peers (Allen et al., 2011; Smith et al., 2005). Such methods will increase instructional interest and will lessen student disconnect from content.

The goal of PBLs is to engage students using problems within a meaningful context (Baeten et al., 2010). Students are provided information with gaps; the learning targets are identified, but students can locate missing information. Students locate research and literature to complete the problem.

Studies show that students who receive PBL instruction have positive long-term learning and significant retention more so than recall-based learning strategies (Dochy et al., 2003; Strobel & Van Barneveld, 2009). PBL situations are deliberately postured to provide missing information that is needed, therefore shifting learning from a lecture style to a style of active engagement.

Presenting the goals for this kind of strategy as well-defined and clear is considered best practice. This strategy is intended to encourage students to conduct self-motivated research (Allen et al., 2011). In a well-constructed PBL classroom, teachers rely on authentic assessments where students are able to demonstrate learning through product development, a performance task, or basic problem solving. For example, science students may complete a unit focused on bacteria where the problem introduced could be the reduction of student absences due to illnesses. Students could learn about bacteria and then develop a video about preventing disease through proper hand washing and knowledge of germs. A performance task could also be a debate, seminar, position paper, or lab report.

If there is difficulty during PBL instruction implementation, it is suggested that the teacher maintain knowledge of students' progression and assist with tasks in the PBL process. The teacher can devise short activities to address instructional gaps to assist students with challenging concepts or to accentuate various aspects of the content (Allen et al., 2011).

PBL BEST PRACTICES: FINDING SOLUTIONS

The following are best practices for educating CLD students using CRPBL:

- Authentic performance tasks are critical to the development of competencies within students, and real-world context in curriculum can become inseparable in an educational experience (Darling Hammond, (2010).
- Educators can aspire to create opportunities where students become critical thinkers and agents of change (Freire, 1970).

Educators who utilize the strategies of problem-based learning create environments that speak directly to best practices. As stated previously, PBLs provide students with real-world problems that require them to develop solutions through the use of their prior knowledge of the topic and identified learning targets developed by the student and teacher. The students take on the role of scientist, lawyer, or community member who has the responsibility to create a solution.

Within the creation of that solution, students create a product or performance task that provides evidence that the learning targets are met. Instead of student assessment remaining the memorization of facts, as it currently exists in traditional classrooms, students are assessed by the application of their knowledge to the problem. This application, instead of memorization (recall), invokes critical thinking that allows deeper learning of the content, as suggested by Baeten et al. (2010). There is a clear connection between what is suggested as best practices for students from CLD backgrounds and the aims of problem-based learning.

INNOVATIONS IN INSTRUCTION FOR PBL IN STEM

Ideas to change instruction can include integrating higher order thinking skills based on PBL and STEM. Higher-order thinking skills integrated into instruction

develop strong cognitive skills for CLD students. Suggested changes can include ways teachers and educators can implement and develop cognitive skills (Bellanca et al., Fogarty, & Pete, 2012; Bellanca et al., 2013; Dana et al., 2013). Students will need to use these skills when implementing PBL and STEM. STEM can help students gain these skills by being

- coupled to literature within the context of standards;
- integrated in a manner to modify instruction for the varied academic level of students within a given class;
- readily used within muti-tiered systems of support (MTSS) for at-risk students and students with mild disabilities alike;
- promoted as interdisciplinary instruction for various STEM classes;
- measured using various means of assessment (i.e., authentic, curriculum-based, performance) including formative and summative measures within classes;
- readily coupled with high-stakes academic testing data provided about students; and
- tied to a thread of specific disciplines and skills to be taught that target cognitive skills.

We recommend that teachers implement higher order thinking skills (i.e., clarifying, reasoning) at a consistent rate, then teach students to become self-regulating in the use of these various skills. *Explicit instruction* and *inquiry-based learning* concepts can help to increase cognitive skill levels and promote PBL.

Explicit Instruction

Explicit instruction, which couples well with culturally responsive teaching, is a well-researched method of teaching in which instruction is designed to assist students in using background knowledge of a specific topic or concept (Hammond, 2019). The method lends itself well to PBL because it encourages higher level thinking strategies that promote high achievement in content areas. Additionally, explicit instruction encourages students to become active participants during instruction (Archer & Hughes, 2010; Goeke, 2008). This method also adapts well to the merger of smaller portions of instructional units so that connections between the last unit taught and the next unit to be taught can be seen as a continuous whole. Consequently, higher levels of cognitive thinking are stimulated while students are encouraged to apply real-world skills to instruction.

Greene (2019) emphasized some key features of explicit instruction that are applicable to students with disabilities and CLD students alike:

- Students who struggle with attention deficit disorders can become more focused because the methodology uses cues to signal actions on specific material.
- Rivera et al. (2009) studied language and reading interventions for English language learners and English language learners with disabilities. They found that explicit instruction was effective. Learners were not inundated with new language demands and therefore had significant increases in achievement.
- Explicit instruction attributes are structured to blend well with multitiered systems of support (MTSS). Both students with disabilities and those at risk for school failure who need intensive instruction need opportunities for repeated, guided, as well as independent practice.

Greene (2019) provided several steps in the application of this method (Table 2.1) that are applicable to students with disabilities and students considered at risk.

Inquiry-Based Learning

Inquiry-based learning is more than a question-based technique; it uses innovative questioning to activate curiosity (Grade Power Learning, 2018; Wolpert-Gowrun, 2016). Inquiry-based learning, a real-life, practical application, positions questions, problems, or scenarios to activate cognitive processes throughout a lesson. This approach improves comprehension and engagement. There are methods that are more student directed and those that are more teacher directed.

A teacher-directed approach begins with information-gathering questions to determine how much students know about the topic. Events that connect to the STEM content for that lesson or unit can be considered. For example, ask students the following:

- "Have there ever been any pandemics like COVID-19?" Films and video clips to provide background information can be provided (e.g., resources about the 1918 pandemic).
- "What is virulent? And why is COVID-19 more virulent than other viruses?"
- "Why did people need to build memorials to Civil War soldiers? Why are these offensive to some cultures?"

Table 2.1. Applying Explicit Instruction

Stage 1: Purpose	• Provide the goals, purpose, and standards. • Link the material to the past lesson/unit. • Explain the skill and the learning outcome.
Stage 2: Model	• Model the technique. • Focus on the most important part of the lesson. • Use a "talk out loud/think out loud" (self-talk) approach as you model.
Stage 3: Arrangement	• Divide the task into specific steps and scaffold material. • The teacher and the student work through an example together.
Step 4: Examples and practice	• Provide multiple opportunities to practice. • Provide guided practice, and then independent practice only after the student understands guided practice. After independent practices are understood, the student is encouraged to generalize. • Analyze student data to delineate needed assistance. • Use graphic organizers and mnemonic memory techniques to ground material learned. • Provide resources, support, and strategies that meld with this approach. • Give written, oral, and group responses. • Encourage ideas.
Stage 5: Feedback	• Provide formative assessments. • Reflect on strengths and challenges. • Provide summative assessments. • Analyze student data. • Expect 80% mastery.

Figure 2.1. Students Can Ask Themselves These Questions

1. Which parts of the lesson are new and what parts do I understand?
2. Is there some manner in which I connect or fit to the lesson or unit?
3. Is this fact or opinion? Is it personal or impartial?
4. If it is personal, does any portion of the subject or objectives suggest any thoughts about something I already know or understand?
5. Is this idea important to my family, community, and the world? To others? Why or why not?
6. What role do I play regarding this topic? Is this part of life?
7. Do I have any questions?
8. Who can I contact or collaborate with to have some input on this topic or idea?
9. How does this really interest or relate to me? Can I create something new with this information?

Heick (2020) suggested several questions students can ask themselves when learning new content (Figure 2.1). Each of these questions can connect to various STEM disciplines and previous and future learning with real-world examples.

The use of questioning sometimes creates challenges. In a research study on questioning in STEM classrooms, Eshach et al. (2014) found that questions provoking higher level thinking are less frequently provided; in 11 out of the 17 instances, in a typical class, *lower order questions* are usually asked. Consequently, teachers increase the frequency of required higher level thinking responses.

Cotton (2001) suggests several tips for teacher proficiency in using inquiry strategies:

- Use a range of questioning, including divergent and high- to middle- and low-level cognitive questions.
- Prewrite questions as part of the unit or lesson before presenting the topic to the students.
- Use higher order cognitive questions in conjunction with grade-level questions during lessons or units' discussions and instruction.
- Use technological apps to pose questions and to foster motivation for those students at lower cognitive levels to increase the participation of students of all levels. This technique protects anonymity, while providing questions to the entire class or

group allows for input from all students. Many times, students are motivated because they find out they know more than they previously believed about the lesson.

Formatively, throughout the lesson, questions provide insight to teachers on how well the student understands and integrates new material and skills. Questions posed at the end of the lesson allow instructors to hear students share their perceptions of cognitive constructions and associations to determine if they need more guided, intermediate, or advanced practice with the concepts or if they are ready to generalize the information to other topics or subjects.

Student-Directed Inquiry-Based Learning Application

Student-directed inquiry-based learning can also be used to develop student-directed projects and therefore helps students become active learners. In other words, students are in charge of their learning, and they take ownership when they are inspired and engaged. When using student-directed strategies, teachers' flexibility is necessary. Still, we suggest that teachers direct the inquiry toward further learning to keep students focused on the goals and objectives of the project. Barnett (2019), Edutopia (2015), and Wolper-Gawron (2016) share student-directed steps in inquiry-based learning:

Step 1: Students develop a questionnaire related to a problem they want to solve. Students can work in groups to determine a problem statement. Internet resources can be used as basic research tools to develop a response. In this step, students make observations.
Step 2: The teacher acts as a guide to help students research and to provide additional resources for students to examine.
Step 3: Students make predictions and form hypotheses.
Step 4: Students collect data and analyze findings.
Step 5: Students present what they learned in a creative presentation (e.g., a website, YouTube, etc.)
Step 6: Students reflect on the learning process.

With student-directed or teacher-directed inquiry-based methods, clear communication between teachers and students is paramount. It is important that learning objectives are discussed at the start and end of the unit or lesson.

CULTURALLY RESPONSIVE TEACHING

Gloria Ladson-Billings, in *The Dreamkeepers* (2009), provides a framework for the culturally relevant teacher. The author expresses the notion that culturally relevant teachers emphasize the strengths and prior knowledge of students from culturally and linguistically diverse (CLD) backgrounds by using the connections between home and school culture. Ladson-Billings (2009) emphasizes the positive effect teachers have when including the perspectives of their students and their community. These culturally relevant teachers have a supportive environment in the classroom so that all students can in some way succeed.

Culturally relevant teaching bridges students' home and school lives while promoting the validity of the knowledge students bring to the table. One of the keys to CLD students' success is a relationship that involves in-depth knowledge of students. Students will develop more of a commitment to learning if they observe the commitment of the teacher.

Dei (1997) in his study contends that a large percentage of student dropouts were unable to connect to any learning in the curriculum because it did not include anything relative to their experiences. Lisa Delpit (2012) finds that the connection to home, school, and community is most important for effective teachers so that they can continuously validate students through these types of connections.

CLD students learn more effectively when their educational experiences include the contributions of "their people" to world progress and scientific development. So, to evoke students' full potential, opportunities need to be provided to explore their roots and discover who they are (Akbar, 1999). Freire (1970) believes culturally responsive teachers awaken the oppressed, and when the oppressed are educated, there is an awakening that retains students in educational settings.

Darling-Hammond (2010) argues that districts develop ways of fostering culturally responsive innovations without sacrificing equity, which provides a free and adequate education for all students. The impact of multicultural education and culturally responsive pedagogy is invaluable in halting the failings of the educational system for CLD children.

It is recommended that teachers work to create culturally responsive lessons linked to learning standards to increase student achievement. Lisa Delpit (2012) challenges teachers to ensure that instruction reflects high standards of learning.

Concepts of Culturally Relevant Pedagogy

There are three concepts of culturally relevant pedagogy (Ladson-Billings & Tate, 1995): (a) academic achievement, (b) cultural competence, and (c) sociopolitical consciousness. These concepts foster success and support pedagogy that allows teachers to help students understand why they are learning. Cultural competence is the teacher's ability to help foster student learning about themselves, others, and how the world functions around that knowledge. This concept helps students understand the culture of power in our society and how to navigate through challenges. Students then see themselves as active participants and can develop a deeper knowledge of self (Milner, 2012). Sociopolitical consciousness centers on the local community and students' families and experiences. Figure 2.2 presents the positive outcomes of culturally relevant pedagogy.

Figure 2.2. Positive Outcomes of Culturally Relevant Pedagogy

Empowers students to do the following:

- Scrutinize educational content and processes
- Generate and build their own meanings
- Flourish in academic and social settings
- Observe inequalities in the local and larger context of the world

Integrates student culture in the following:

- Instruction in the learning setting
- A focus on students' knowledge of background and world
- The positive effect of being part of the non-majority culture

Fashions classroom frameworks that

- are stimulating and creative;
- produce a focus on student academic achievement;
- shape cultural competence; and
- connect the curriculum and instruction to real-life situations.

Student Voice Using Culturally Responsive Instruction

When a student's voice or input is included in the culture of a school, it encourages the student to engage in the learning process (Jackson, 2011). The opportunity for student expression helps engage students in authentic interactions with teachers that lends to positive changes in curriculum, assessment, and significant instruction that shape schooling and the classroom. Students are then treated as competent co-creators and collaborators of knowledge rather than empty vessels teachers fill with knowledge. Students' voices encourage active participation in their learning (Friere, 1970).

School Inequities, Opportunity, Achievement, and Educational Gaps

Schools prepare approximately 20% of their students for problem-solving and analytical work in advanced or honors courses. In the recent past, these courses have been typically unavailable to African American, Latino, and Native American students but are more available today. These groups typically received curriculum focused on vocational and basic skill instruction. Though districts have been working to address disproportionality, more efforts are encouraged. Unfortunately for many, the goals of an inclusive curriculum and equal access to a quality education have remained elusive. Darling-Hammond (2010) contended that lack of teacher knowledge intensifies inequalities in schools. These unprepared teachers are usually placed in districts where there are CLD students and students from low socioeconomic backgrounds, thereby exacerbating the achievement gaps. Perhaps examining the type of instruction and services can shed more light on the achievement gap in schools. Lisa Delpit (2012) emphasizes that CLD children are not born with deficits that contribute to the achievement gap.

Biography-Driven Instructions for CLD

As described and explained by Herrera (2016) the heart of biography-driven instruction (BDI) involves getting to know students holistically in order to be truly responsive to their cultural and linguistic assets and needs. Educators understand the sociocultural, linguistic, cognitive, and academic dimensions of the CLD student biography (Herrera, 2016; Herrera & Murry, 2011) to answer the following types of questions:

- Sociocultural: What brings students life, laughter, and love?
- Linguistic: In what ways do students use their first language (L1) and second language (L2) for comprehension, communication, and expression?
- Cognitive: How do students know, think, and apply?
- Academic: To what degree do students have access, engagement, and hope?

Students with Mild Disabilities

The U.S. Department of Education (2010) concluded that African American and Latinx students (6–21) were 1.5 times more likely to receive special education services as compared to those in all other racial groups combined. While African American and Latinx children make up only 17% of the public school population in the United States, they make up 41% of the special education population (Kozol, 2005). Low employment and low socioeconomic status, heightened arrest, and "prison tracks" are cited as culminations of special education for non-majority populations (King, 2005; Losen & Wellner, 2001). Houchins and Shippen (2012) described the correlation between the history of special education and youth in the juvenile justice system. Roughly 40% of incarcerated youth have disabilities compared with approximately 12% of students in public schools (Gagnon et al., 2009).

CULTURALLY RESPONSIVE PBL (CRPBL)

The call to incorporate culturally responsive teaching with that of problem-based learning is vital in the effort to increase student engagement. Linda Darling-Hammond (1997) urges educators to teach students with realistic roles by engaging them in authentic educational opportunities through the development of performance tasks (e.g., any learning activity or assessment that asks students to perform to demonstrate their knowledge, understanding, and proficiency). Problem-based learning experiences encourage the use of performance tasks that will allow students to apply what they have learned rather than regurgitate information. The educator is taking their knowledge of the students and selecting PBL problems that are relevant to students' cultural capital.

In a case study developed by Timothy Berry (2013) that explored CRPBL, the findings spoke directly to best practices necessary to effectively educate CLD students. As a result of the PBL, students felt comfortable with CLD and participating. Students reported feeling empowered and unafraid to share their perspectives with the teacher. This study suggested that CLD were successful because they were engaged in the learning process. In addition, Berry argued that student choice in the selection of performance tasks and the topic of diversity were key factors in the successful increase of engagement. Berry (2013) continued by stating that CRT should be at the forefront of student learning for Black males in order to make the topic meaningful.

SUMMARY: RECOMMENDATIONS FOR TEACHERS

When teachers develop what is to be learned and then assessed, they begin to use innovative ways to entice students to participate. Producing or allowing a question to drive inquiry is good practice. The driving question can be one that will motivate students to take appropriate steps in their research, activities, and tasks.

It is strongly recommended that teachers are acquainted and have a relationship with the students in order to create a CRPBL environment. Knowledge of students and their capabilities helps with planning related STEM tasks and activities, which is an important aspect of creating a CRPBL.

Alignment to standards. It is central to align STEM projects, topics, activities, and tasks to state standards to assure tasks are consequential exercises. It is important that teachers remember an increase in student engagement increases student academic opportunities, and formative, summative, as well as high-stakes measures can be considered.

Entry event. The entry event can be so compelling and innovative that students attach themselves to the problem. We recommend that teachers introduce resources (e.g., video, letter, speaker, play, etc.) to emphasize aspects of cultural responsiveness (e.g., familiar terms, neighborhood, images, etc.) as well as problem-based strategies.

Next steps. The problem can be designed to highlight its authenticity. It is favorable when teachers find (a) what students know and (b) what students need to know. The teacher can provide leading questions and record student responses under each category. The teacher can then use this information to match responses to grade-level content and standards to determine what the students need to learn. Questioning can then focus on what students need to learn and channel energy toward solving the problem, which can be defined in the unit or lessons.

Assessment/product. Student assessment can include a product or performance task, such as an experiment, position paper, multimedia presentation, the development of a program, or several other resources for students to show achievement. It is important that students understand measures such as rubrics for what was learned and produce evidence of rigor that exhibits the quality of the product.

This chapter examined the concept of PBL as a strategy to provide students with relevant, engaging CRT that will raise student achievement. This approach will be referred to as CRPBL. The use of multiple instructional strategies such as inquiry-based learning and explicit instruction can be contributing factors to increase student engagement and achievement.

REFERENCES

Akbar, N. (1998). *Know thy self*. Mind Productions & Associates, Inc.

Allen, D. E., Donham, R. S., & Bernhardt, S. A. (2011). Problem-based learning. *New Directions for Teaching and Learning, 128*, 21–29.

Archer, A. L., & Hughes, C. (2010). *Explicit instruction: Effective and efficient teaching (what works for special-needs learners)*. Guilford.

Baeten, M., Kyndt, E., Struyven, K., & Dochy, F. (2010). Using student centered learning environments to stimulate deep approaches to learning: Factors encouraging or discouraging their effectiveness. *Educational Research Review, 5*(3), 243–260.

Barnett, S. (2019, July 31). *Using scientific pedagogy to teach history*. George Lucas Educational Foundation. https://projectdragonfly.miamioh.edu/wp-content/uploads/2019/10/Barnett_Shana_Edutopia_July2019.pdf

Barrows, H. S. (1996). Problem-based learning in medicine and beyond: A brief overview. *New Directions for Teaching and Learning, 1996*(68), 3–12. http://dx.doi.org/10.1002/tl.37219966804

Bellanca, J. A., Fogarty, R. J., & Pete, B. M. (2012). *How to teach thinking skills within the Common Core: 7 key student proficiencies of the new national standards*. Solution Tree Press.

Bellanca, J. A., Fogarty, R. J., Pete, B. M., & Stinson, R. L. (2013). *School leaders guide to common core*. Solution Tree Press.

Berry, T. A. (2013). *Contextual pedagogy: A praxis engaging Black male high school students toward eliminating the achievement gap* [Doctoral dissertation, Minnesota State University, Mankato]. https://cornerstone.lib.mnsu.edu/etds/81/

Cotton, K. (2001). *Classroom questioning*. North West Regional Educational Laboratory. https://educationnorthwest.org/sites/default/files/resources/classroom-questioning-508.pdf

Dana, N. F., Burns, J. B., & Wolkenhauer, R. (2013). *Inquiring into the common core*. SAGE.

Darling-Hammond, L. (1997). *Doing what matters most: Investing in quality teaching*. National Commission on Teaching and America's Future.

Darling-Hammond, L. (2010). *The flat world and education: How America's commitment to equity will determine our future*. Teachers College Press.

Dei, G. J. (1997). *Reconstructing "drop-out": A critical ethnography of the dynamics of Black students' disengagement from school*. University of Toronto Press.

Delpit, L. (2012). *Multiplication is for White people: Raising expectations for other people's children*. The New Press.

Dochy, F., Segers, M., Van den Bossche, P., & Gijbels, D. (2003). Effects of problem-based learning: A meta-analysis. *Learning and Instruction, 13*(5), 533–568.

Edutopia. (2015, August 24). *Inquiry based learning: Harnessing student's curiosity to drive learning*. https://www.edutopia.org/practice/wildwood-inquiry-based-learning-developing-student-driven-questions#:~:text=Inquiry%2Dbased%20learning%2C%20rather%20than,agency%20and%20critical%20thinking%20skills

Eshach, H., Dor-Ziderman, Y., & Yefroimsky, Y. (2014). Question asking in the science classroom: Teacher attitudes and practices. *Journal of Science Education and Technology, 23*(1), 67–81.

Freire, P. (1970). *Pedagogy of the oppressed*. Continuum.

Gagnon, J. C., Barber, B. R., Van Loan, C., & Leone, P. E. (2009). Juvenile correctional schools: Characteristics and approaches to curriculum. *Education and Treatment of Children, 32*(4), 673–696.

Gay, G. (2000). *Culturally responsive teaching: Theory, research, and practice*. Teachers College Press.

Goeke, J. (2008). *Explicit instruction: A framework for meaningful direct teaching*. Pearson.

Grade Power Learning. (2018, April 03). *What is inquiry-based learning (and how is it effective)?* https://gradepowerlearning.com/what-is-inquiry-based-learning/

Greene, K. (2019). *Explicit instruction: What you need to know*. Understood. https://www.understood.org/en/school-learning/for-educators/universal-design-for-learning/what-is-explicit-instruction

Hammond, L. (2019). Explainer: What is explicit instruction and how does it help children learn? *The Conversation*. https://theconversation.com/explainer-what-is-explicit-instruction-and-how-does-it-help-children-learn-115144

Harry, B., & Klingner, J. K. (2006). *Why are so many minority students in special education?: Understand race & disability in schools*. Teachers College Press.

Heick, T. (2020, May 2018). *15 questions students can ask themselves when learning new content*. Teach Thought. https://www.teachthought.com/critical-thinking/15-questions-help-students-respond-to-new-ideas/

Herrera, S. (2016). *Biography-driven culturally responsive teaching*. Teachers College Press.

Herrera, S., & Murry, K. (2011). *Mastering ESL and bilingual methods: Differentiated instruction for culturally and linguistically diverse (CLD) students* (2nd ed.). Allyn & Bacon.

Houchins, D. E., & Shippen, M. E. (2012). Welcome to a special issue about the school-to-prison pipeline: The pathway to modern institutionalization. *Teacher Education and Special Education, 3*(4), 265–270.

Jackson, Y. (2011). *The pedagogy of confidence: Inspiring high intellectual performance in urban schools*. Teachers College Press.

King, J. E. (Ed.). (2005). *Black education: A transformative research and action agenda for the new century*. Erlbaum.

Kozol, J. (2005). *The shame of the nation: The restoration of apartheid schooling in America*. Three Rivers Press.

Krashen, S. D. (1984/2002). Bilingual education and second language acquisition theory. In C. F. Leyba (Ed.), *Schooling and language minority students: A theoretical framework* (2nd ed., pp. 47–75). Legal Books.

Ladson-Billings, G. (2009). *The dreamkeepers: Successful teachers of African American children* (2nd ed.). Jossey-Bass.

Ladson-Billings, G., & Tate, W. (1995). Toward a critical race theory of education. *Teachers College Record, 97*(1), 47–68.

Linnenbrink, E. A. (2007). The role of affect in student learning: A multi-dimensional approach to considering the interaction of affect, motivation, and engagement. In P. A. & P. Reinhard (Eds.), *Emotion in education* (pp. 107–124). Elsevier Academic Press.

Losen, D., & Welner, K. (2001). Disabling discrimination in our public schools: Comprehensive legal challenges to inappropriate and inadequate special education services for minority students. *Civil Liberties Law Review, 36*(2), 407–260.

Loyens, S., Jones, S., Mikkers, J., & Gog, T. (2015). Problem-based learning as a facilitator of conceptual change. *Learning and Instruction, 38*(10), 34–42.

Marzano, R. J., Gaddy, B. B., & Dean, C. (2000). *What works in classroom instruction*. Mid-Continent Research for Education and Learning.

Milner H. R. (2011). Culturally relevant pedagogy in a diverse urban classroom. *Urban Review, 43*, 66–89.

Moll, L. C., Amanti, C., Neff, D., & Gonzalez, N. (1992). Funds of knowledge for teaching: Using a qualitative approach to connect homes and classrooms. *Theory into Practice, 31*(2), 132–141.

Rivera, M. O., Moughamian, A. C., Lesaux, N. K., & Francis, D. J. (2008). *Language and reading interventions for English language learners and English language learners with disabilities.* RMC Research Corporation, Center on Instruction.

Shabatura, J. (2020). *Using Bloom's taxonomy to write effective learning objectives.* https://tips.uark.edu/using-blooms-taxonomy/

Smith, K. A., Sheppard, S. D., Johnson, D. W., & Johnson, R. T. (2005). Pedagogies of engagement: Classroom-based practices. *Journal of Engineering Education, 94*(1), 87–102.

Sousa, D. A. (2006). *How the brain learns* (3rd ed.). Corwin.

Strobel, J., & Van Barneveld, A. (2009). When is PBL more effective? A meta-synthesis of meta-analysis comparing PBL to conventional classrooms. *Interdisciplinary Journal of Problem-Based Learning, 3*(1), 44–58.

Tomlinson, C. A. (2001). *How to differentiate instruction in mixed-ability classrooms* (2nd ed.). Association for Supervision and Curriculum Development.

U.S. Department of Education. (2010). *29th annual report to Congress on the implementation of the Individuals with Disabilities Education Act, 2007.* https://files.eric.ed.gov/fulltext/ED516249.pdf

Vygotsky, L. S. (1978). *Mind in society: The development of higher psychological processes* (M. Cole, V. John-Steiner, S. Scribner, & E. Souberman, Eds.). Harvard University Press.

Wolpert-Gawron, H. (2016). *Inquiry-based learning: What the heck is inquiry-based learning? Teachers use inquiry-based learning to boost student engagement.* Edutopia. https://www.edutopia.org/blog/what-heck-inquiry-based-learning-heather-wolpert-gawron

CHAPTER 3

Selection of Reading Materials, Technology, and STEM Activities

Diane Rodriguez, Jugnu Agrawal, Justin Coles, and Gary Hoag

Cultural and linguistically diverse groups (CLD), people with disabilities, and women are severely underrepresented in many science, technology, engineering, and mathematics (STEM) fields (Wright, 2018). Educators can help change that discrepancy as young people exhibit significant interest in career paths in STEM. As this trend continues, how can educators link the STEM fields with literature? The connection between literature and STEM needs ingenuity, and this chapter attempts to foster that in a viable, stimulating way using lessons, ideas, and examples based on popular young adult literature. Our goal is to provide lesson ideas.

BENEFITS OF CONNECTING LITERATURE AND STEM

Researchers have found numerous benefits of combining literature and STEM studies. The most important of these are gaining a desire for gender equality, problem solving, curiosity about STEM, and elements of reading comprehension. Wright (2018) investigated the impact of nonfiction, engineering-centric literature on student attitudes toward engineering and found that infusing literature into STEM studies has multiple benefits. In particular, students develop more positive attitudes toward gender equality and an increased presence of young women in STEM careers result. Importantly, students' problem-solving abilities also increased. Consistent with Wright's (2018) work, Mahzoon-Hagheghi and colleagues (2018) also found that science-themed literature increases students' curiosity in STEM as well as improves reading, cognitive ability, reasoning, remembering, and critical thought.

Mahzoon-Hagheghi and colleagues (2018) note that by "incorporating this type of literature, teachers can introduce different contexts, concepts, and

cultures that can initiate discussion about a science topic" (p. 42). They ascertain that using science literature books with young adults in the classroom environment not only helps them make connections to their world, but also encourages them to develop reading comprehension. Indeed, numerous studies have shown that when teachers integrate STEM and literature, student performance scores increase along with their interest and enthusiasm for science (Mahzoon-Hagheghi et al., 2018). These results reveal the educational value of literature to learning in STEM.

In particular, reading comprehension appears to improve when using literature in the STEM classroom. Based on student knowledge, and with appropriate teaching support, teachers can help students create specific connections between the literature and science topics. Teachers can also engage students in a discussion about those connections (guided learning). Alternatively, teachers can model how they interpret passages in literature based on their knowledge of STEM and request student input (modeling). In both examples, young adults benefit from opportunities to relate their background knowledge to what they have read in order to make personal connections between literature and STEM concepts. For adolescents with disabilities and English learners (ELs), direct instruction on interpretations of literature and STEM is crucial to understanding the content. We recommend teachers use a variety of strategies, such as questioning, visualizing, inferring, determining importance, making connections, and synthesizing. Use of those instructional strategies will reinforce content knowledge in STEM for students.

The current goal of education is to make learning accessible to *all* students. Integrating literature into STEM benefits all students, including adolescents from culturally and linguistically diverse (CLD) backgrounds, students at multiple reading levels, second language learners, students with disabilities, and students with various learning preferences. Research shows that students will be more capable of understanding STEM concepts when literature is integrated with instruction in STEM (Honey et al., 2014).

USING LITERATURE TO TEACH STEM IN THE INTERDISCIPLINARY CLASSROOM: SELECTING A BOOK

We believe an interdisciplinary approach is better than focusing on a single subject. Student response to high-interest texts (Tankersley, 2005) indicates less concern with the specific class in which they are enrolled and much more concern about the STEM content. According to Lott and colleagues (2013), "STEM-related activities have become more common in schools [due to] the hope of generating greater student interest in STEM" (p. 65). If we wish to reverse the relative absence of CLD students, those with disabilities, those whose first language is not English, or women in STEM, educators must ensure that literature is used in STEM classes. This section attempts to stimulate curriculum development to connect literature and STEM.

To easily connect literature to STEM, Mayer (1995) provided the following list of questions for teachers:

1. Is the science concept recognizable?
2. Is the story factual?
3. Is fact discernible from fiction?
4. Does the book contain misrepresentations?
5. Are the illustrations accurate?
6. Are the characters portrayed with gender equity?
7. Are animals portrayed realistically?
8. Is the passage of time referenced adequately?
9. Does the story promote a positive attitude toward science and technology?
10. Will students read or listen to the book?

Goodreads (2020) identifies some of the most-read young adult books, which fit most of Mayer's criteria. This partial list includes *The Hunger Games* (Collins, 2008), *Twilight* (Meyer, 2005), *Harry Potter and the Philosopher's Stone* (Rowling, 1997), the *Percy Jackson and the Olympians* series (Riordan, 2005–2009), *The Maze Runner* (Dashner, 2009), *Holes* (Sachar, 1998), *To Kill a Mockingbird* (Lee, 1960), and *Hidden Figures* (Shetterly, 2016) as well as others. Readability might pose a challenge for young adults with disabilities, but students' interest level—if instructors appropriately construct lessons—will be high. Using films that accompany each book and other suggested strategies can motivate students to explore STEM concepts.

Table 3.1 offers books that may be used to connect literature to STEM content. For each book, themes and STEM topics are listed. The texts cover an array of topics and are written by authors from CLD backgrounds.

The texts in Table 3.1 provide examples to promote text integration, which sparks student interest to engage in STEM study. For example, many students

Table 3.1. Adolescent Literature Pertaining to STEM Content and Themes

Literature	Book Themes	Technology and STEM Content
The Hunger Games (Collins, 2008)	Survival, innovation	Technology, engineering and design
Children of Blood and Bone (Legacy of Orisha) (Adeyemi, 2018)	Chemistry, math, physics	Metaphysics, science and mysticism
Opposite of Always, (Reynolds, 2019)	Time travel	Physics
Twilight (Meyer, 2005)	Shape shifting and vampires	Physics
The *Harry Potter* series (Rowling, 1997)	Magic	Chemistry, metaphysics
American Street (Zoboi, 2017)	Mysticism (alchemy) and the supernatural	Chemistry and medicine
The Marrow Thieves (Dimaline, 2017)	Global warming, the human body	Climate change, immune system
Cinder (Meyer, 2012)	Technology	Earth and space, disease, biomechatronics (robotics, neuroscience)
Every Day (Levithan, 2013)	Science, physics	Gender fluidity, body transformations, metaphysics
The *Maze Runner* series (Dashner, 2009)	Mathematics, logics	Engineering, cryptology, kinesiology, mathematics
Holes (Sachar, 1998)	Geology	Geology (earth structure), sediments
Hidden Figures (Shetterly, 2016)	Technology	Engineering, mathematics

may have encountered the *Harry Potter* book series but may not have had the educational opportunity to link their deep interest in the books to a budding interest in chemistry. Though some students have an interest in particular content areas, they may not believe they can follow those career paths. The teaching of literature and STEM encourages students.

Teaching Strategies and Activities

Educators seek to increase literacy skills by teaching students to apply knowledge through critical analysis and by stimulating higher order cognitive skills. Memorizing facts and information is important, but the application of critical thinking skills increases one's ability to analyze and solve problems. Drawing on literature, teachers expose students to cognitive progressions through (a) knowledge and comprehension; (b) application and analysis; (c) synthesis; and (d) evaluation. In order to develop cognitive reflective skills, we must engage students in a variety of activities. For example, students can describe experimental designs and define key terminology in their own words. Additionally, students can create graphic organizers or diagrams to summarize key elements. Further, students can assess the effectiveness of analytical tools. Students can also generate new hypotheses based on the readings and propose additional experiments. Moreover, students can communicate (i.e., orally, written, via technology), using a number of avenues, what they have read and learned.

Science is taught to increase skills in observation, prediction, and hypothesizing in order to acquire facts, concepts, and principles in the natural universe. A holistic approach to STEM topics through a literary lens engages young adults and provides opportunities to apply scientific knowledge, which may make students more amenable to learning in science. Students experience the excitement of discovery. In the context of STEM instruction, think of literature as a vehicle to teach students STEM facts, concepts, procedures, and principles. Teachers can use strategies to develop literacy skills and to help students better understand science. Skills and confidence gained acquiring STEM content may pivot students toward a career in STEM, whether engineering, data science, or medicine, for example. Literacy skills connected to science are valuable for enhancing young adults' understanding of scientific inquiry. Teachers can engage students in inquiry using any combination of the following strategies (Cox, 2012):

- Making observations
- Asking questions

- Reading books and discovering old and new knowledge
- Planning an investigation
- Using tools to gather, organize, analyze, and interpret data and information
- Making predictions and suggesting answers and explanations
- Communicating findings and results to others

Many students struggle with math because they suffer from a lack of academic success, which creates a sense of failure. According to Furner (2018), teachers can use literature to reach students by reading the text to them, which will help students acquire math concepts without being intimidated. Furner (2018) discusses many benefits of using literature to teach mathematics concepts: Teachers can (a) include math concepts in the context of a story; (b) combine integrated studies with reading, writing, speaking, and listening; (c) advance thinking skills; (d) permit a variety of responses; (e) allow for historical, cultural, and practical application and connectivity; (f) promote the use of math manipulatives as they relate to the story; (g) evaluate a student's understanding by reading and questioning; (h) use a wide range of books that are connected with mathematics concepts; and (i) use problem solving and active involvement from the context of the story (Furner, 2018). Such instructional activities make for an enjoyable experience for both students and the teachers.

Contemporary students use technology, but they fail to understand that technology can be a product of engineering. Teachers must leverage students' technical skills and let them explore and develop (engineer) innovative ways to learn with the latest technology. STEM instructional activities complement rather than supplement traditional literature courses. Conversely, with appropriate instructional support, students in literature courses may reinforce their knowledge of STEM topics. According to Stone and colleagues (2008), time spent on math and science concepts via applied learning can boost achievement.

Universal Design for Learning (UDL) encourages teachers to use multiple means of expression, representation, and engagement (Gronneberg & Johnston, 2015). UDL promotes engagement by helping students make connections using a variety of teacher-led and peer-assisted learning strategies (PALS). McLeskey (2017) has shown that "[a]ctive student engagement is critical to academic success" (p. 84).

Cooperative learning structures help increase engagement by fostering interaction among students (Kagan, 2009). In Kagan structures, instructional

Figure 3.1. Morpheme Card

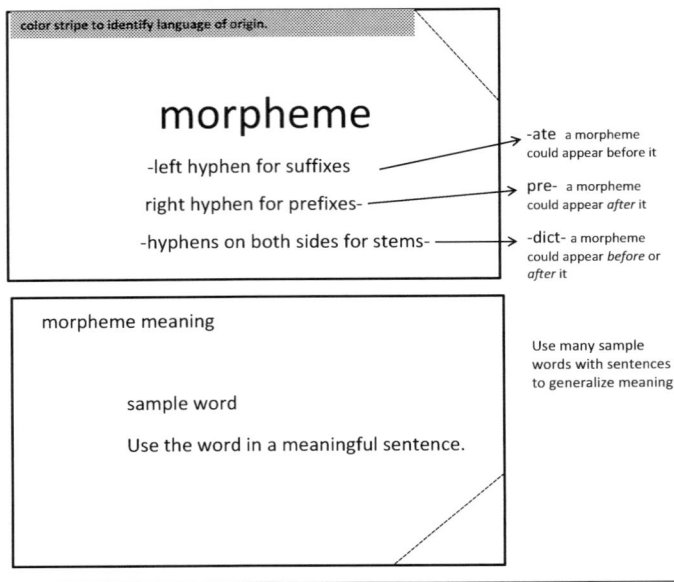

strategies designed to promote cooperation and communication are used across content areas and grade levels. There are more than 200 Kagan structures from which teachers can choose to match a specific activity to promote academic learning. For example, teachers can require students to master vocabulary and ask or require students to create morpheme cards, which identify the morpheme (word part; see Figure 3.1) on the front and the meaning and use of the word in a meaningful sentence on the back. Teachers appoint a leader to move in and out of collaborative groups and the group compares notes.

Vocabulary

Vocabulary is a particular issue of concern for students with learning disabilities, CLD students, young women, and students whose first language is not English. Morphemes are the smallest unit of language with meaning. In order to teach vocabulary, exercises that cause students to focus on the morphemes in a word (morphemes bibliography cards), its use in one or more meaningful sentences, and/or its representation with graphical means are most likely to encourage engagement (Sedita, 2009). The two types of vocabulary most at issue are literary terms and content-specific vocabulary.

Students can master literary terms by using the story map, a device that provides supporting information and it integrates maps, legends, text, photos, and video that helps students explore this content.

Morpheme cards, various foldables, and graphic organizers can permit students to become engaged with content vocabulary. Graphic organizers such as the "Words Alive" map (Figure 3.2) will allow students to focus on words in literature, guess their meaning, paraphrase (i.e., find their own definition) from a dictionary definition, create an image, find related words, and use the words in meaningful sentences. Such multiple, multisensory inputs tend to result in mastery of the word. A blank copy of the "Words Alive" map can be downloaded from the Virginia Department of Education website (www.doe.virginia.gov/instruction /english/middle/wordsalive/index.shtml)

Making Global Connections

STEM education through literature has achieved a global presence by increasing the number of CLD students using literature in STEM education, which encourages those students to examine gender and racial equality in all cultures. Kamkwamba and colleagues' (2012) solutions emphasize STEM regardless of age or cultural background, and establish it as an inclusive topic in global communities. As educators explore ways to increase the number of young people in the STEM fields, paying particular attention to CLD learners in the United States will help meet the needs of all students. Students must engage in hands-on activities that build their interest, knowledge, and self-efficacy in STEM (Ralston et al., 2013). Since unfamiliarity with English may pose considerable challenges for CLD adolescents, hands-on activities are crucial to learning in the STEM field. Indeed, Means and colleagues (2017) concluded that STEM research opportunities and project-based instructional approaches increase CLD students' interest in STEM. According to Means et al. (2017), schools should integrate STEM subjects into literature instruction to increase real-world STEM experiences.

Technology plays a substantial role in promoting collaboration, communication, and engagement in order to find solutions to problems. A student-driven environment with participatory learning encourages students to take risks, test, innovate, and communicate ideas. Identifying literature that will help provide opportunities for these connections is crucial.

Assessment

Assessment is a key part of planning instruction and effective teaching. Special education teachers should use multiple sources of performance assessment.

Figure 3.2. Words Alive Map

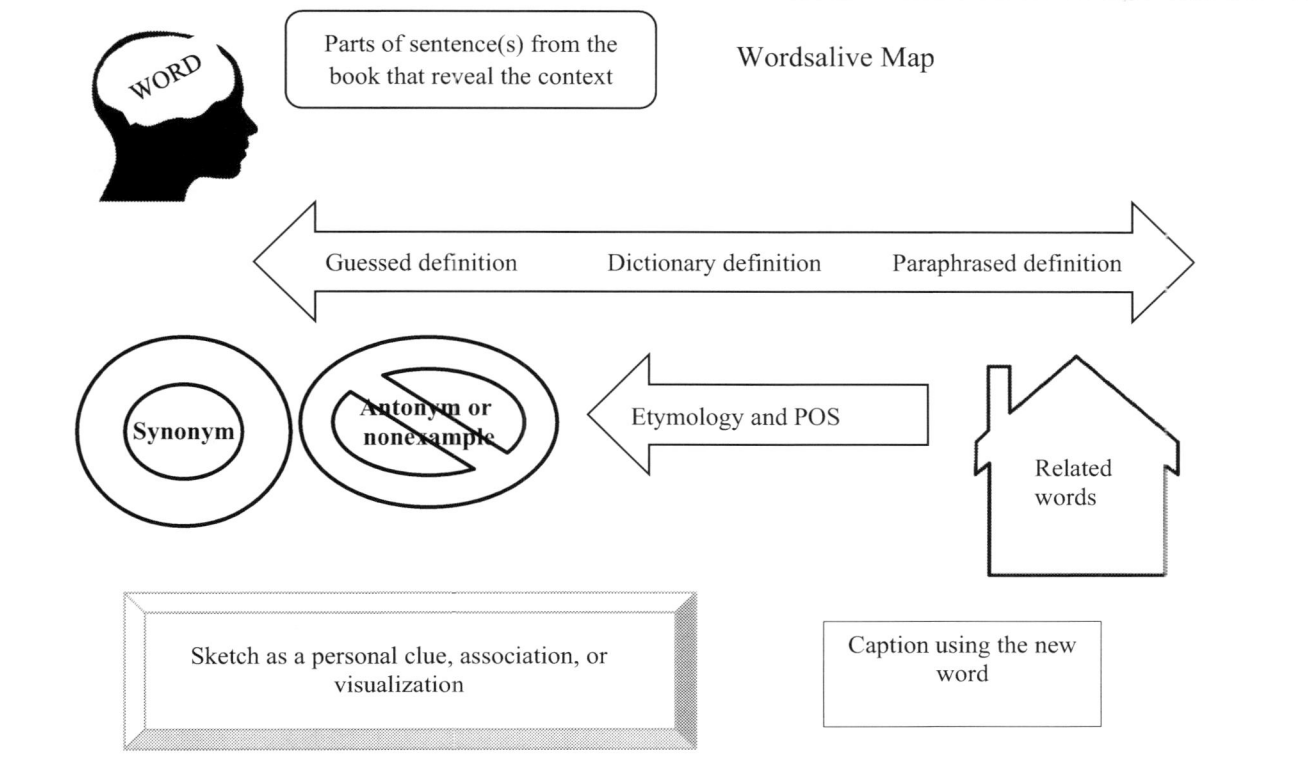

Teachers can use performance-based measures to ensure that students have acquired the targeted skill. Such assessments allow multiple presentation formats in order to increase student learning and give clear expectations for grading through a well-developed rubric (McLeskey et al., 2017).

Formative assessments provide ongoing feedback to teachers and students. Educators can use different types of formative assessments, including student self-reflections, checklists, assignments, observations, discussion boards, projects, and journals. Teachers can use data from the assessments to guide their teaching and review concepts as needed. Summative assessments have evolved over time and are performance based, such as portfolios (both online and/or paper), final projects, presentations, and paper-pencil tests.

Technology tools are useful to develop embedded real-time feedback. According to the National Education Technology Plan (NETP) update report, "Assessments delivered using technology can provide a more complete and nuanced picture of student needs, interests, and abilities than traditional assessments, allowing educators to personalize learning"

(Office of Educational Technology, 2021). Educators can use digital tools, such as Google forms, WebAssign, Quizlet, and Blackboard, Canvas, or another learning management system (LMS), to design assessments. Technology tools also allow educators to scaffold and differentiate assessments for students with disabilities.

Using Common Core Standards

Common Core State Standards (CCSS) outline the learning goals a student should master. These standards were developed to provide consistency across states to prepare students with 21st-century skills to be successful in college, career, and life. Forty-one states, the District of Columbia, and four territories initially adopted these academic standards for public schools. While many states have withdrawn from CCSS, for these and other states using local academic standards, Common Core standards correlate and are available on most state Department of Education websites. Common Core standards are used in conjunction with integrating STEM and literature. There are specifics as to how you infuse STEM with literature.

Step 1. Select the book, unit, or set of chapters you want to discuss from the classroom informational text and determine what skills you want to teach. For instance, you want to teach biology vocabulary. What skill, subskill, or other standard will be embraced? For example, the *Twilight* Saga is a series of four vampire-themed fantasy romance novels that tell about the life of Bella, a teenage girl who falls in love with a vampire named Edward. Students can view a clip from the first Twilight movie, "Twilight Biology Class Scene Edward's Golden Eyes," and list and define the science vocabulary words mentioned in the video clip, as well as other words from their informational text. Additionally, they can discuss cold- and warm-blooded animals, bats and their habitat, and animals that drink blood. This book encourages careers in biology as a phlebotomist, blood splatter expert in criminology, and other blood sciences.

Always use interrelated skills such as vocabulary and its use, which includes *writing and literacy* that emphasizes the use of (a) domain-specific vocabulary; (b) specialized reference materials (e.g., dictionaries, glossaries, and thesauruses); (c) figurative language, word relationships, and nuances in word meanings; and (d) different mediums (e.g., print or digital text, video, multimedia) to present an idea.

Step 2: You can emphasize culture. It is also important to use a text with diverse themes that can relate to urban culture. *Twilight* emphasizes a set of vampires that differ from the human population. Differences in race, ethnicity, age, and so on can be discussed in relation to the majority population.

Step 3: Decide what teaching strategies and activities can be used. Present a high-energy, engaging activity related to STEM. Several different teaching strategies and methodologies can be outlined and suggested. The activities will be infused using the literature.

- Activities can include (a) listing the science vocabulary words mentioned in the video clip; (b) defining the words; (c) listing the parts of speech; (d) selecting the varied word meanings, including Greek/Latin affixes/suffixes/prefixes and the inferred meaning of the words; and (e) virtual flashcards.
- Discussions can include (a) cold- and warm-blooded animals, (b) bats and their habitat, (c) animals that drink blood, (d) the legends of vampires, and (e) who Bram Stoker is.

- Role-play activities can include (a) acting out meanings or various words and (b) writing a skit using various biology words.

Step 4: Assessment. Measures can include a rubric that assesses several related activities. The scoring can include a number of varied types of measures that can tap into different talents students may possess. Measures can include (a) enunciation and pronunciation; (b) writing skills; (c) skit, roleplay, or video performance activities; and (d) the use of vocabulary words in context.

Step 5: Generalize the skills to other literature books. Determine how the knowledge of the meaning of the vocabulary words can assist in the next lesson, in the community, and at home.

USING HIGH-LEVERAGE PRACTICES AND STRATEGIES FOR STUDENTS WITH DISABILITIES

High-leverage practices (HLPs) in special education cover four major areas: collaboration, assessment, social/emotional/behavioral practices, and instruction. Instructional strategies cover the majority of HLPs and are crucial for providing effective instruction for students with disabilities (McLeskey et al., 2017). Teachers can select and teach essential curriculum components, identify essential prerequisites and foundational skills, and assess student performance to monitor learning. Several strategies, such as pre-teaching vocabulary within the context of a text/task, using leveled texts, providing audio supports (read aloud and audiobooks), inferring, making connections, questioning and visualizing, can help to increase reading comprehension. Visual supports, such as graphic organizers, pictures and photographs, real-life objects and manipulatives, charts, tables, and timelines, can make the text more concrete for students receiving special education services. Additionally, strategies such as explicit modeling, peer tutoring, cooperative learning strategies, chunking information, gradually releasing information, teacher feedback, and providing wait time for responses will help to actively engage students in the learning process (Berry, 2006). The lessons in this book are designed for general and special education students. There are many strategies and accommodations built in the lessons for at risk and/or students with disabilities, but some students may need additional supports.

SUPPORTING LEARNING IN THE VIRTUAL ENVIRONMENT/DISTANCE LEARNING

Education's response to virtual learning has changed teaching and learning in unprecedented ways in the year 2020. The amount of time that students are expected to spend on instructional activities varies widely, and teachers are implementing creative ways to engage students in virtual learning. Adapting to a new reality in teaching, virtual learning might become a new norm in the educational environment. Therefore, understanding concepts and ideas to support learners to learn remotely might seem overwhelming. The good news is that there are opportunities to improve distance learning through professional development and teacher preparation programs. In addition, when students have teachers' support and guidance to connect with their virtual learning classes, they are more likely to enjoy their literature/STEM lessons.

Multiple factors can be considered while engaging students in online learning. Technology tools and the online platforms used to communicate with the students are extremely important. Several platforms such as Google Meet, Zoom, Blackboard, and Microsoft Teams are available for teachers to meet with their students (Basilaia & Kvavadze, 2020). Additionally, the United Nations Educational, Scientific, and Cultural Organization (UNESCO) provides a comprehensive list of distance learning solutions. Just like in the brick-and-mortar classroom, teachers can start with establishing norms for the virtual classroom. This can include keeping cameras on, utilizing protocols to ask/answer questions using the hand-raising feature, and being respectful and engaged.

Virtual instruction can include synchronous and asynchronous activities or can be a hybrid. Synchronous activities are led by the teacher in real time. During synchronous activities, teachers can engage students using the interactive features/tools of the platform. For example, chat, online polls, emojis, interactive whiteboards, Jamboard, and Padlet are few of the many tools available to engage students. Additionally, teachers can use breakout rooms to group students for collaborative work, which can include working on a project; documenting their thinking using Google Docs, which can be shared with the class later (Minero, 2020); and providing feedback to each other. Many of the cooperative learning strategies can be adapted for the virtual environment, such as virtual gallery walks, think-pair-share, and virtual discussion boards.

For asynchronous activities, teachers can use the several G Suite applications such as Google Classroom, Google Docs, and Google Forms. Google Forms can also be used to create graded assessments. Teachers can provide the option for students to complete a quiz once or multiple times depending on the purpose of the quiz.

Providing accommodations in the virtual environment can sometimes be easier than in the actual classroom as many online platforms have built-in accessibility features. For clarifying directions, teachers can provide video of the directions, screencast how to complete a task, and create "how to" guides with visual supports. Students can also use online checklists for assignments and tasks for each week such as Google Keep, the Notes app, bulleting in Google Docs or Microsoft Word, and Todoist. For visual supports, teachers can provide digital copies of the documents to support the instructional activity, information sheets on how to use a tool, and editable documents of visual supports. Students can use tools such as Visor or Liner on the Chrome browser and Immersive Reader in Microsoft products for visual tracking while reading. Keyboarding and speech-to-text features can be utilized for supporting students with writing. Closed captioning can be done in Google Slides, PowerPoint, and Google Meet (FDLRS Administration Project, n.d.). These are just a few examples of accommodations that can be used to ensure that remote learning is accessible. Some practical examples from Herrmann (2020), Minero (2020), and Taplin (2020) that can assist with problem-based learning are listed:

- To discuss or work on a project collectively, students can meet in small groups. A designated recorder and discussion leader are chosen. The group cooperatively can complete a diagram (e.g., Venn) and convene to discuss with the larger class.
- For group projects, teachers can provide an outline and list of materials. Students will be able to gather materials individually but can work communally as teams through virtual avenues to build models or use problem-based learning techniques.
- The chat feature can be used to check for student understanding. Students can type "T" or "F" for true-or-false questions or can provide individual answers to questions.
- Students can critique, comment, or analyze varied lessons, formats, or strategies using feedback forms emailed to the instructor that

highlight student difficulty. They can reply to simplistic questions such as "What is one thing you found helpful?" and/or "What is one thing that could be improved?"

- For complex projects, online forums provide students a way to have back-and-forth discussions to problem solve or have discussions among and in between groups.
- Virtual gallery walks provide students an opportunity to view each other's projects while stimulating ideas for new connections and problem-solving methods. Additionally, students can present ideas through screencast or other web-illustrated methodologies and publications (discussed further in Chapter 7).
- Particularly for upcoming projects, teachers can provide a video accompanied with a written schedule succinctly summarizing projects for the week, including a list of materials, goals, and objectives of the task. The video can be accessible for students and caregivers.
- A 3- to 5-minute clip of the goals and objectives and key points of various lessons can provide students and caregivers previews of upcoming lessons and reviews of previously taught lessons and assignment due dates.
- Locating and using technological tools (Chapter 7) on a consistent basis can be helpful. Students and caregivers can learn how to use these tools efficiently and can be a part of each lesson. Introduce new tools one at a time so that students and caregivers can become familiar with their use.
- In a virtual settings, "Try It, Talk It, Color It, Check It" is a process that can be helpful for teamwork. Students first attempt to independently solve a problem. If they have not solved the problem in a given time (e.g., 2 to 5 minutes) they are assigned to a group to work through the problem in the talk it phase. The color it phase is illustrated through set colors that indicate that (a) the problem is solved, (b) additional assistance is needed, or (c) the group is ready to help other students if need be. The check it phase involves meeting with the larger group to discuss and share elements of the problem.
- Student can be provided a problem to solve independently. This independent "try it" work time is essential to give students an opportunity to generate their own ideas and/ or questions about the problem.
- When time was up, we assigned students to breakout rooms with a social and emotional learning (SEL)–focused prompt (e.g., "What makes you feel happiest?"). The prompts we selected, examples of "listening circles," were supported by CASEL's research on helping students develop their social awareness and relationship skills.

LESSON PLANNING

The mnemonic **ATTACK** is the framework for organizing lessons: **a**pplication, **t**eaching strategies, **t**rial/ try-out, **a**ssessment, **c**ognitive reflection, and **k**eep, retain, generalize. A lesson plan template is provided so that you see the whole lesson at a glance (see Figure 3.3).

Application. The application section presents the standards for the lesson or unit and can be related to one subject area or focus on multiple ones. Teachers can present the lesson goals and objectives in persuasive, imaginative, and clear language.

Teaching strategy. Presentation of the skill in an engaging manner is key to motivating students. Each teaching strategy should include the following four components:

- *Introduction*: Ideas are presented to stimulate interest by using media (Scheibe & Rogow, 2008). Educators should introduce the literature piece to the students by explaining the book (i.e., author, producer, setting up the context, and making connections with the educational content), guiding student learning through questions, showing movie segments, and allowing students to demonstrate learning.
- *Materials*: Essential materials needed to teach the lesson should be listed in this section.
- *Time*: Approximate time needed to teach the lesson or the unit should be clearly stated.
- *Essential vocabulary*: The required vocabulary words from the STEM informational text or selected activity are listed and taught, especially for students with disabilities to build background knowledge.

Figure 3.3. Lesson Plan Template

Lesson Plan/Target:

Application: Long/Short-Term Goals/Objective(s):	**Standard:** **Learning Target:**		
Teaching Strategy (Opening):	**Introduction:** **Materials:** **Essential Vocabulary:**		
Trial/Try-out: If using multiple standards, identify the standard number in parenthesis next to the activity	Focus 1:	Focus 2:	Focus 3:
Assessment/Progress monitoring			
Cognitive Reflection:			
Know, Retain, Generalize			
Notes			

Trial/try-out. The trial or try-out section should include activities that require application of the targeted skill:

- *Technology*: Internet, video, and other technologically related strategies that focus on the goals, objectives, or learning target (i.e., standards) can be provided.
- *Additional exercises*: Scaffolds and additional supports for accessibility to students with disabilities are presented here.

Assessment. Multi-presentation formats are presented in this section to measure assignments. Student projects or assessments are tied to the goals, objectives, and learning targets.

Cognitive reflection. Researchers have repeatedly shown that metacognition—the ability to think

intentionally about processes—is one of the keys to learning. Combining literature and STEM encourages students to think about how they learn. Analyzing routines can be used to support cognitive reflections among students (Project Zero, n.d.).

Keep, retain, generalize. The keep, retain, generalize section can include guiding questions that help students use the information in a different setting or use the information in a varied manner that connects skill with other subject areas.

SUMMARY

Researchers can examine the use of literature as an integrated part of STEM programs in secondary schools. Teachers can integrate investigations and literature to study both topics together while making them accessible to CLD students, students with disabilities, and students who are ELs. The ultimate goal is for consideration of STEM topics to become commonplace in discussions of literature and part of the everyday curriculum for adolescents. After all, use of relevant literature in the teaching of STEM subjects helps lower math anxiety, pique student interest, and increase confidence to succeed in the STEM fields. Teachers who integrate literature in STEM instruction for adolescent students are able to nurture their interest and understanding of STEM. STEM topics invite critical thinking, which requires students to move beyond consumption of knowledge and into the analysis, evaluation, and synthesis of ideas. Some students will be inspired to pursue a career path in a STEM field. Adolescents who choose paths outside STEM will still be able to appreciate the skills they have learned and may transfer the disciplined methods of STEM subjects into other fields of study, as well as apply them in their daily lives.

REFERENCES

Adeyemi, T. (2018). *Children of blood and bone (Legacy of Orisha)*. Henry Holt and Company.

Basilaia, G., & Kvavadze, D. (2020). Transition to online education in schools during a SARS-CoV-2 coronavirus (COVID-19) pandemic in Georgia. *Pedagogical Research*, *5*(4), em0060. https://doi.org/10.29333/pr/7937

Berry, R. A. (2006). Teacher talk during whole-class lessons: Engagement strategies to support the verbal participation of students with learning disabilities. *Learning Disabilities Research & Practice*, *21*(4), 211–232. https://doi.org/10.1111/j.1540-5826.2006.00219.x

Collins, S. (2008). *The hunger games*. Scholastic Press.

Cox, C. (2012). *Literature-based teaching in the content areas*. SAGE.

Dashner, J. (2009). *The maze runner*. Delacorte Press.

Dimaline, C. (2017). *The marrow thieves*. Cormorant Books.

FDLRS Administration Project. (n.d.). *Providing accommodations in a virtual environment*. https://andoveres.ocps.net/UserFiles/Servers/Server_55270/Image/PAVE%20(002)%20(1).pdf

Furner, J. M. (2018). Using children's literature to teach mathematics: An effective vehicle in a STEM world. *European Journal of STEM Education*, *3*(3), 14. https://doi.org/10.20897/ejsteme/3874

Goodreads. (2020). *Best books of 2020*. https://www.goodreads.com/choiceawards/best-books-2020

Gronneberg, J., & Johnston, S. (2015). *7 things you should know about universal design for learning*. Educause Learning Initiative. http://www.educause.edu/library/resources/7-things-you-should-know-about-universal-design-learning

Herrmann, Z. (2020). *How to make the most of student feedback during distance learning*. Edutopia. https://www.edutopia.org/article/how-asking-more-effective-questions-can-increase-student-learning-math

Honey, M., Pearson, G., & Schweingruber, H. (2014). *STEM integration in K–12 education: Status, prospects, and an agenda for research*. National Academies Press. https://doi.org/10.17226/18612

Kagan, S. (2009). *Kagan structures: A miracle of active engagement*. https://www.kaganonline.com/free_articles/dr_spencer_kagan/281/Kagan-Structures-A-Miracle-of-Active-Engagement

Kamkwamba, W., Mealer, B., & Zunon, E. (2012). *The boy who harnessed the wind*. Dial Books for Young Readers.

Lee, H. (1960). *To kill a mockingbird*. Lippincott.

Levithan, D. (2013). *Every day*. Ember.

Lott, K., Wallin, M., Roghaar, D., & Price, Y. (2013). Engineering encounters: Catch me if you can! *Science and Children*, *51*(4), 65–69.

Mahzoon-Hagheghi, M., Yebra, R., Johnson, R. D., & Sohn, L. N. (2018). Fostering a greater understanding of science in the classroom through children's literature. *Texas Journal of Literacy Education*, *6*(1), 41–50.

Mayer, D. (1995). How can we best use literature in teaching children's science concepts? *Science & Children*, *32*(6), 16–19, 43.

McLeskey, J., Barringer, M.-D., Billingsley, B., Brownell, M., Jackson, D., Kennedy, M., Lewis, T., Maheady, L., Rodriguez, J., Scheeler, M. C., Winn, J., & Ziegler, D. (2017). *High-leverage practices in special education*. Council for Exceptional Children & CEEDAR Center.

Means, B., Wang, H., Wei, X., Lynch, S., Peters, V., Young, V., & Allen, C. (2017) Expanding STEM opportunities through inclusive STEM-focused high schools. *Science Education*, *101*(5), 681–715. https://doi.org/10.1002/sce.21281

Meyer, M. (2012). *Cinder*. Feiwel and Friends.

Meyer, S. (2005). *Twilight*. Little, Brown and Company.

Minero, E. (2020, August 21). *8 strategies to improve participation in your virtual classroom*. Edutopia. https://www.edutopia.org/article/8-strategies-improve-participation-your-virtual-classroom

National Association for Media Literacy Education. (2007). *The core principles of media literacy education*. https://namle.net/publications/core-principles/

National Research Council. (2014). *STEM integration in K–12 education: Status, prospects, and an agenda for*

research. National Academies Press. https://doi
.org/10.17226/18612

Office of Educational Technology. (2021). *National education technology plan*. https://lincs.ed.gov/professional
-development/resource-collections/profile-902

Project Zero. (n.d.). *Project Zero's thinking routines toolbox*.
Harvard Graduate School of Education. https://
pz.harvard.edu/thinking-routines#CoreThinking
Routines

Ralston, P. S., Hieb, J. L., & Rivoli, G. (2013). Partnerships
and experience in building STEM pipelines. *Journal
of Professional Issues in Engineering Education & Practice*,
139(2), 156–162. https://doi.org/10.1061/(ASCE)
EI.1943-5541.0000138

Reynolds, J. A. (2019). *Opposite of always*. Katherine Tegen
Books.

Rowling, J. K. (1997). *Harry Potter and the philosopher's
stone*. Bloomsbury.

Sachar, L. (1998). *Holes*. Farrar, Straus and Giroux.

Scheibe, S., & Rogow, F. (2008). *12 Basic ways to integrate
media literacy and critical thinking* (3rd ed). Project
LookSharp.

Sedita, J. (2009). *The key vocabulary routine*. Keys to Literacy.

Shetterly, M. L. (2016). *Hidden figures: The American dream
and the untold story of the Black women mathematicians
who helped win the space race*. William and Morrow.

Stone, J. R. I., Alfeld, C., & Pearson, D. (2008). Rigor and
relevance: Enhancing high school students' math
skills through career and technical education. *American Educational Research Journal, 45,* 767–795. https://
doi.org/10.3102/0002831208317460

Taplin, A. (2020). Adapting an effective math collaboration activity for distance learning. *Edutopia*. https://
www.edutopia.org/article/adapting-effective-math
-collaboration-activity-distance-learning

Tankersley, K. (2005). *Literacy strategies for grades 4–12:
Reinforcing the threads of reading*. Association for
Supervision and Curriculum Development.

United Nations Educational, Scientific, and Cultural Organization. (n.d.). *Distance learning solutions*. https://
en.unesco.org/covid19/educationresponse/solutions

Wright, G. A. (2018). Engineering attitudes: An investigation of the effect of literature on student attitudes
toward engineering. *International Journal of Technology and Design Education, 28,* 653–665. https://doi
.org/10.1007/s10798-017-9417-0

Zoboi, I. (2017) *American street*. Balzer and Bray.

CHAPTER 4

Science and Literature

Gloria Campbell-Whatley, Kim Reddig, and Deondra Gladney

Literature and science are intertwined; yes, they are different, but entwined. Historically, science and literature are linked together (Vlahakis et al., 2014). Literature and science use different techniques, but both provide pleasure and adventure, can increase knowledge, and provide unique experiences. For example, in the era of the COVID-19 pandemic, scientists reminded the general population about previous historical and literary writings about pandemics. They spoke of attempts to create vaccines and other methods and strategies to slow or stop the spread of a deadly virus.

There are other perspectives where science and literature link. Adolescents are curious about both of these topics, and as they study them, imagination and innovation, as well as research and investigation, are inspired. Culture is full of movies and novels that present a plethora of scientific experiences from the past that are now popular. The *Twilight Zone* series, *Star Trek* series, *Terminator* series, *Frankenstein* remakes, and *Amazing Stories* remakes have been rewritten, updated, and can be streamed through many sources such as Netflix, Roku, and Hulu. As referred to earlier in this book, there are a number of books that have been produced as movies or series (e.g., *Twilight, Percy and the Olympian*). These books are powered by the connection between science and literature. Integrating science and literature and teaching and learning encourages various scientific techniques (Erduran, 2020).

SCIENCE AND LITERATURE FOR ADOLESCENTS

Adolescents' perceptions of science differ between gender, creativity and nurturing, awareness of scientific knowledge, and the objective nature of science (Emran et al., 2020). All of these differences relate to cognitive styles and motivation (Zeyer, 2018). Attitudes, interest, and the utility of science play a large part in adolescents' learning (Henriksen et al., 2015).

DeWitt and Archer (2015) found that adolescents girls had difficulty envisioning scientific careers for themselves more often than adolescent boys. The researchers found that for science, girls need to be engaged in "identity work" to develop aspirations and maintain them overtime. Girls need high achievement, exposure, and a positive mindset in STEM in the earlier grades before high school. Girls focus on practical applications of science whereas boys focus new knowledge, theory, and logic; therefore, teacher lesson designs need to focus on the varied scientific concepts (Flegg & Burke, 1995; Tsai & Liu, 2005). Reilly and colleagues (2019) found that boys and girls equally have the talent and ability to engage in scientific research provided the right set of motivation, encouragement, and experiences.

Teachers need the skills to inspire a wide range of student knowledge, including low- to middle-class socioeconomic status, the knowledge of various cultures, and a wide range of scientific activities (DeWitt & Archer, 2015). These skills also include high-leverage instructional practices for at-risk students and students with special needs (i.e., students with disabilities, gifted students) such as providing long- and short-term goals systematically designed to increase performance toward learning targets. In addition, adapting the curriculum to fit basic scientific tasks by differentiating assignments according to disability status, gender, and cultural learning styles is desirable (McLeskey et al., 2017).

CAREERS

Science includes a wide range of career choices. These areas can be introduced to adolescents to let them know what the fields of science can offer them.

Biochemists and biophysicists. Biochemists and biophysicists study the chemical and physical principles of living things and biological processes, such as cell development, growth, heredity, and disease. Additionally, they handle various technologies in their work, such as lasers and microscopes, to conduct scientific experiments and analyses. Typically, they use X-rays and computer modeling software to observe the structures of proteins and molecules and research chemical enzymes to synthesize DNA. In their research they do the following:

- Perform and conduct complex projects
- Direct laboratory collaborative groups that may include others in the STEM field
- Research the effects of drugs, hormones, and biological processes
- Review literature and the findings of researchers to formulate technical reports or make recommendations for the health of the community from their findings
- Secure funding such as grants

Chemists. Chemists observe matter by examining atoms and molecules and their magnitude and reaction rates to understand unfamiliar substances. Additionally, they create new compounds that can be used in everyday life. Their common duties and responsibilities include the following:

- Creating, synthesizing, measuring, testing, analyzing, and providing quality control for new substances and compounds and providing data on chemical or physical properties and the effectiveness of varied and new applications
- Performing lab work to develop or test products or formulations
- Creating models and theories and examine their uses
- Analyzing compounds to determine composition, structure, relationships, or reactions
- Helping solve quality issues and troubleshoot manufacturing issues
- Developing new products and improving existing products and manufacturing processes
- Writing technical papers or reports while working with others in STEM fields
- Evaluating laboratory safety procedures and ensure compliance
- Preparing test solutions and compounds for laboratory testing
- Communicating with customers and suppliers to determine what they require from a product

Conservationists. Conservationists manage outdoor environments, including parks, forests, and the like. They examine soil and water to find avenues for land use without harming the environment. They perform many tasks, including the following:

- Educating society about the environment by providing expertise on the national forest
- Connecting with the government, industry groups, and other large institutions to teach them to conserve

Ecologists. Ecologists focus on the relationships between the organisms of the natural world and their environments. Their practices are extensions of other STEM careers and include biology, earth science, physics, and chemistry. Their careers include conducting green research by working for the following:

- Federal institutions such as the Environmental Protection Agency (EPA)
- Private companies to work on green policy formation
- Lobbyists to advocate for groups to launch green campaigns
- Corporations and the government, as well as the public, to make informed decisions regarding the use of natural resources

Forensic scientists. Forensic scientists aid in criminal investigations by collecting and analyzing evidence. They specialize in either crime scene investigation or laboratory analysis. The tasks or responsibilities of a forensic scientist are as follows:

- Collecting and analyzing evidence and communicating the material for law enforcement
- Traveling to the scene of the crime to collect evidence themselves

Geoscientists. Geoscientists study the physical aspects of the Earth and its composition, structure, and processes, as well as pollution, food webs, paleontology, habitats, and climate change. The study involves all of Earth's dynamic processes, not just those on the crust. In their task they learn about the following:

- The Earth's past, present, and future
- Minerals, mountains, rocks, resources, eruptions, erosion, sediment, caves, and maps

Hydrologists. Hydrologists measure the use and height of water at any given time. They predict flooding and check water flow and precipitation. In performing task on their jobs, they do the following:

- Measure water temperature to help determine the source and predict future weather patterns
- Collect information about changes in ground height
- Deduct information about moisture content of the soil to possibly predict drought patterns and assist farmers with crop growth
- Measure the pH of soil or water in an environment to determine unsafe organisms

TEACHER APPLICATION TIPS

Mesci and colleagues (2020) mentioned several factors of the pedagogical content of science that teachers need to know:

- Have a strong knowledge of the science curriculum. The teacher must be prepared to do the following:
  - » Identify core concepts, modify activities, and know connections between science and STEM concepts
  - » Know how to analyze formative and summative assessments
- Understand students and know the differences in various cultures. The teacher must be prepared to do the following:
  - » Connect culturally
  - » Design tasks related to interest, diversity in learning styles, motivation, need, and developmental level
  - » Understand learning difficulties (i.e., motivations, special education or at-risk categories)
- Determine a number of evidence-based instructional strategies such as the following:
  - » Explicit and inquiry-based instruction
  - » Examples, models, and varied activities

Table 4.1. Summary of Literature Books/Films and Related Skills

Book	Summary and Science Skill
Lesson 4.1 *Twilight* vampire series	The story of a teenaged vampires. Scientific terms in biology are the focus of the lesson.
Lesson 4.2 *Immortal Life of Henrietta Lack*	The story is about an African American woman whose cells are taken from her without permission to become important cancer research. The lesson explores cytology and presents ethical issues of using someone's cells without permission.
Lesson 4.3 *The Boy Who Harnessed the Wind*	This story is about a boy who was able to build a windmill out of rudimentary equipment. The lesson focuses on activities related to the properties of water and its impact on food and drinking conditions.
Lesson 4.4 *Percy Jackson and the Lighting Thief*	The book is about a half-god teen and other mythological creatures in a camp. The pros and cons of drugs are discussed and several bases of Greek, Roman, and Egyptian mythology.
Lesson 4.5 *Ender's Game*	Ender, a teen, is used to destroy enemies of Earth in space. Gravity, weight, and mass are discussed. Genocide and its effect on various cultures are discussed.
Lesson 4.6 *After Earth*	A man and his son's spaceship crashes on Earth. The story centers around their survival on a hostile planet. The lessons center on pheromones and their uses and the respiratory systems.
Lesson 4.7 *Gattaca*	The world is divided between those who have been genetically altered and those who have not. The lesson focuses on genetics, DNA, and discrimination.
Lesson 4.8 *The Matrix*	Computer programmers find that they are alive only through a computer construct and the machines have overtaken the Earth. Lessons center on alternative fuel sources to save energy.
Lesson 4.9 *Holes*	A group of boys are being punished at a camp by digging holes. Desert plant and animal life are discussed.
Lesson 4.10 *Star Wars: Splinter of the Mind's Eye*	Luke Skywalker and Princess Leia are marooned together on a strange planet where they encounter struggles against the forces of the evil Galactic Empire and Darth Vader. Assistive technology for persons with disabilities are discussed.

- Provide varied means of assessment (e.g., presentations, drawings, discussions, lab reports, and observations)
- Teach science concepts and subject matter

SUMMARY OF LESSONS

It is important that teachers use science to engage adolescents and foster the use of complicated texts such as classics like *Frankenstein* or *Beowulf* and determine how these stories can be used with today's youth. For example, both of these stories contain monsters; *Beowulf* contains dragons and *Frankenstein* includes man-made monsters. They include depictions of war and zombies similar to those seen in the video games of today (Schwartzbach-Kang & Kang, 2018). These stories can be incorporated into lessons to promote various STEM areas such as chemistry, anatomy, and physiology for *Frankenstein*, and *Beowulf* can be used to introduce modern-day dragons like the Komodo dragon.

The lessons in this chapter include several well-known books and/or movies that are listed in Table 4.1.

LESSON 4.1. SCIENCE VOCABULARY

The *Twilight* Saga, Book One, by Stephenie Meyer (2005)

Summary. The *Twilight* Saga is a series of four vampire-themed fantasy romance novels that tell about the life of Bella, a teenage girl who falls in love with a vampire named Edward. High-school student Bella Swan, always a bit of a misfit, doesn't expect life to change much when she moves from sunny Arizona to rainy Washington state. Then she meets Edward Cullen, a handsome but mysterious teen whose eyes seem to peer directly into her soul. Edward is a vampire whose family does not drink blood, and Bella, far from being frightened, enters a dangerous romance with her immortal soulmate.

Application

Goal. To help students with vocabulary in informational STEM text.

Objective. Students will be able to pronounce and define at least 20 STEM-related words out of 25.

Standards. To view the Common Core Standards that correspond with this lesson, please visit the *STEAM Meets Story* page on www.tcpress.com and click on the Resources tab.

Teaching Strategy

Introduction. Students will view the following clip from the first *Twilight* movie, "Twilight Biology Class Scene Edward's Golden Eyes" (https://www.youtube .com/watch?v=Vc1UqeHhjeo) (*Handout 4.1A-Vocabulary Words*) (Literacy L.6–10.4, L.6–10.4.a, L.6–10.5, L.6–10.5.b, L.6–10.6, L. 6–10.4.b, L. 6–10.4.b, L. 6–10.4.c, L. 6–10.4.c, L.6–10.5.8; Literacy and Technology RL. 6–10.7, Science and Technology RST, 6–8.4).

a. List the science vocabulary words mentioned in the video clip.
b. Define words.
c. List the parts of speech.

Handout 4.1A. Vocabulary Words

Word	Meaning(s)	Part of Speech	Use in Context	Suffixes and Prefixes	Inferred Meaning
WORDS FROM THE VIDEO CLIP					
WORDS FROM THE INFORMATIONAL TEXT					

d. Are there varied word meanings?

e. What are the Greek/Latin affixes/suffixes/prefixes?

f. What are the inferred meanings of the words (check dictionaries and definitions online)?

g. Are there conflicts and differences in the inferred meanings?

Materials

- Handouts
- Selected clips
- Computers or tablets
- Copy of *Twilight* the book
- Informational text

Time. One to two class periods

Essential vocabulary. Look at the captions and the words in bold in the assigned informational text chapter. What words do you need to study and define? Use in context? Define? Spell? (Writing WHST, 6–10.2.d; Literacy L.6–10.6) (*Handout 4.1A*). *Note*: The handouts presented here are simplified due to limited space. To view the original full-color handouts, which can be downloaded and printed for classroom use, please visit the *STEAM Meets Story* page on www.tcpress.com and click on the Resources tab.

Trial/Try-Out

Students will be able to create a paragraph that illustrates and conveys the meaning of the words listed

from the informational text (*Handout 4.1B*) (Literacy L.6–10.4.a, L.6–10.4.b, L.6–10.4.c, L.6–10.4.d.).

Technology. Students will write an interactive scene and role-play or act out and film the meaning of vocabulary words used in the activity or informational text (*Handout 4.1B*) (Literacy and Technology RL. 6–10.7 and Reading Informational Text RI. 6–10.7).

Additional Suggestions

1. Students can make virtual flash cards using the WADE mnemonic to remember words and meanings. These can be posted on a vocabulary website. (Literacy and Technology R.L. 6–10.7 and

Handout 4.1B. Write a Paragraph or an Interactive Scene

Reading Informational Text RI. 6–10.7) (Literacy L.6–10.5.b, L.6–10.4.6, L.6–10.5.b) (*Handout 4.1C*).

   a. **W**rite the word and definition on a virtual flashcard.

   b. **A**rticulate the word using the dictionary pronunciation key.

   c. **D**raw or find a digital a picture that reflects the meaning of the word.

   d. **E**valuate word usage by using it in a sentence.

2. Read and discuss *Twilight* (Writing WHST. 6–10.2.d) (Literacy 6–10.4.a, L.6–8.5.b) (Literacy and Technology RL. 6–10.7, R.I. 6–10.7).

   a. Discussions can include (a) cold- and warm-blooded animals, (b) bats and their habitat, and (c) animals that drink blood.

   b. Discuss: Are vampires real or did the stories arrive from legend?

   c. Discuss: Who is Bram Stoker?

Handout 4.1C. Virtual Flash Card

Write the word	Microscope: An optical instrument having a magnifying lens or a combination of lenses for inspecting objects too small to be seen or too small to be seen distinctly and in detail without aid
Article the word	Mahy-kruh-skohp
Draw or find a picture	
Evaluate word usage	You can see some microbes under a microscope

Assessment

Students will be evaluated using the rubric on *Handout 4.1D* that includes several related activities. The scoring will include (a) enunciation and pronunciation, (b) skit, roleplay, or video and (c) the use of words in context.

Cognitive Reflection

How will the knowledge of the meaning of these words assist you in your next lesson? In the community? At home? (Literacy L.6–10.6).

Keep, Retain, and Generalize

Is there a relationship among words that may be antonyms and synonyms between this lesson and the previous lesson? (Literacy L.6–10.5.b).

Handout 4.1D. Rubric

	Excellent	Good	Need Assistance
Word Pronunciation and Enunciation (10 points)	Clear pronunciation, enunciation, and speed when speaking	Clear pronunciation and enunciation but speaking too fast	Enunciation muddled and words are at a slow tone
Skit, Role-Play, Video (20 points)	Very engaging and interactive	Engaging but not interactive	Neither engaging nor interactive
Word Context (20 points)	Several words used in appropriate context	One word used in appropriate context	No words used in appropriate context

---

LESSON 4.2. SCIENCE ON CELLS

The Immortal Life of Henrietta Lacks by Rebecca Skloot (2017)

Summary. Henrietta Lacks developed cervical cancer around the age of 30 and was treated at Johns Hopkins Hospital. Without her consent or that of her family, doctors removed a small sample of flesh. George Gey, a scientist, was able to use Henrietta's cells to create the first immortal cell line. At the time, scientists had a very difficult time keeping human cells alive in culture. This breakthrough enabled scientists to test the effects of chemicals and new technologies on cells without harming human beings. The cell line, known as HeLa (the first syllables of Henrietta's first and last names) went on to revolutionize science.

Application

Goal(s). To help students do the following:

- Understand the structures and functions of living organisms

- Understand the hazards caused by agents of diseases that affect living organisms
- Summarize the basic characteristics of viruses, bacteria, fungi, and parasites relating to the spread, treatment, and prevention of disease
- Explain the difference between epidemic and pandemic as it relates to the spread, treatment, and prevention of disease
- Understand how biotechnology is used to affect living organisms
- Summarize aspects of biotechnology, including the following:
  » Specific genetic information available
  » Careers
  » Economic benefits to local economies
  » Ethical issues
  » Implications for agriculture

Objectives. Students will be able to do the following:

- Demonstrate an understanding of the ethics and biotechnology behind HeLa cells and how the use of the cells in research has impacted society, families, legislation, and medicine.
- Evaluate the ethics of the use of HeLa cells from different perspectives.

Standards. To view the Common Core Standards that correspond with this lesson, please visit the *STEAM Meets Story* page on www.tcpress.com and click on the Resources tab.

Teaching Strategy

Introduction. Teacher will explain the following:

- What cells are and how they don't normally replicate outside the body
- Definition of biotechnology and its past, current, and possible future roles in society
- Background on the behavior of cancer cells

Materials

- Handouts
- Selected video clips
- Computers
- Copies of excerpts from the book *Immortal Life of Henrietta Lacks*
- Informational audio clip: "Informed Consent and Medical Research"

Time. One to two class periods

Handout 4.2A. Warm-Up Activity

Directions: Create words having to do with cells for each letter of Henrietta Lacks's name	
H	
E	
N	
R	
I	
E	
T	
T	
A	
L	
A	
C	
K	
S	

Essential vocabulary. The list of words for students related to science from *The Immortal Life of Henrietta Lacks* are listed in *Handout 4.2B*. Provide a text or a picture definition demonstrating your understanding of the words. *Handout 4.2C* provides some examples of definitions that are applicable to this lesson.

Trial/Try-Out

1. Students will read "Henrietta Lacks' 'Immortal' Cells" (https://www.smithsonianmag.com/science-nature/henrietta-lacks-immortal-cells-6421299/), an interview with Rebecca Skloot, author of *The Immortal Life of Henrietta Lacks*, that explains how Henrietta Lacks's cells became the first immortal cell line. They will also be provided with excerpts from Skloot's book. Students will be placed in groups of four. Each group will brainstorm words related to cells using the Internet. Then they will write a word having to do with cells for each letter of Henrietta Lacks's name (see *Handout 4.2A*). The words will then be defined and discussed as a group (RST.11–12.6, RST. 9–10.6, RI.6.7, RL.6.7, L.6.4.c, L.7.4.c., L.8.4.c., L.9–10.4.c., L.9–10.4, L. 11–12.4). Students will explore the following guiding questions:
   - How would you feel if this had happened to your family member?
   - Would you donate yourself to science if you didn't get any kind of compensation/recognition?
   - Do you think this could happen today?

Handout 4.2B. Essential Vocabulary

Vocabulary Terms	Text Definition	Picture Definition
Mapping genes		
Cloning		
Raw material		
The common rule		
Biobanks		
Donor restrictions		
Inventive effort		
Genetic rights		
DNA testing		
Genetic disease		

Handout 4.2C. Definitions for Essential Vocabulary

Essential Vocabulary	Definitions
Mapping genes	The process of creating a genetic map to better understand a person's genetic makeup
Cloning	The process of creating an exact genetic copy of a gene, which results in exact copies of every piece of DNA between the two cloned organisms
Raw material	In the case of genetic research, human tissue that has not been modified by researchers. It is tissue that is as it was when removed from the person's body.
The common rule	A federal law that requires scientists to tell people if they are participating in research, that their participation is voluntary, and that they can withdraw from the study at any time, without penalty
Biobanks	Laboratories where tissue from patients is stored for possible research
Donor restrictions	In biomedical research, there are restrictions donors can make on the use of their tissues once removed from their bodies. This issue is highly controversial.
Inventive effort	The process of turning cells into an invention by doing research that modifies the cells in some way
Genetic rights	A person's right to privacy with respect to genetic information and materials
DNA testing	Tests scientists conduct on DNA to extract information
Genetic disease	Diseases that get passed down through generations via parents' genes

2. Students will use the article "Henrietta Lacks' 'Immortal' Cells" (https://www.smithsonianmag.com/science-nature/henrietta-lacks-immortal-cells-6421299/) to answer questions listed on *Handout 4.2D*. Additional questions and resources related to this activity can be found on the ScienceNetLinks website (http://sciencenetlinks.com/lessons/immortal-life-henrietta-lacks/; http://sciencenetlinks.com/student-teacher-sheets/key-concepts-teacher-sheet/).

3. Students will then listen to a news report titled "Informed Consent and Medical Research" (https://www.pbs.org/wnet/religionandethics/2010/06/25/june-25-2010-listen-now/6572/). Discuss the following: Different cultures have different beliefs about health, religion, and death, as depicted in the report. What are some of these differences?

4. Students will form a group of four students. On a T-chart (*Handout 4.2E*) list reasons why families should be compensated if a family member's cells are used for scientific study and a list of reasons why they should not. Have groups share what they listed on the T-chart (*Handout 4.2E*).

5. Students will complete a summative activity (*Handout 4.2F*).

Handout 4.2D. HeLa: Medical Miracle and Ethical Dilemma

Read the article "Henrietta Lacks' 'Immortal' Cells" and answer the following questions:

1. What does the author mean when they say they need cell lines to be "immortal?"
2. Why were Henrietta Lacks' cells important?
3. In your own words define informed consent.
4. Why was there a lot of confusion about HeLa cells throughout the years?
5. In your own words define bioethics.
6. What does ethics mean to you?

Handout 4.2E. T-Chart Activity

Make a T-chart to list reasons families should be compensated if a family member's cells are used for scientific study and a list of reasons why they should not.

Why	Why Not

Handout 4.2F. Summative Activity

Pretend that you are asked to write a section of your last will and testament that details what will happen to your body when you have passed. Explain the terms in which you will allow your body to be used for science and medicine.

- You can choose one of the following three options or create your own terms using a combination of options:
  1. Refuse to donate anything to science and medicine.
  2. Donate any and/or all body parts without the expectation of anything in return.
  3. Donate your body to science and medicine with the expectation that your family will be compensated based upon how it is used.
- You must use at least four terms from the key concepts chart in your statement.
- Consider cloning, stem cell research, and organ donation.
- Be sure to explain if your terms are influenced by religious or other cultural beliefs.
- Your statement should be at least 200 words and written in complete sentences.

Technology. Computers, internet, video clips

Discuss: After reading the materials about Henrietta Lacks, students will pick three scientific and medical breakthroughs that have been made possible as a result of HeLa cells. Explain how HeLa cells were used in each situation.

Handout 4.2G. Rubric for Summative Assignment

Below Expectations	Meets Expectations	Exceed Expectations
Student makes limited use of terminology and demonstrates basic knowledge of content through descriptions and examples.	Student uses considerable and relevant terminology and demonstrates substantial knowledge and understanding of content through descriptions, explanations, and examples.	Student consistently uses a range of terminology accurately and demonstrates detailed knowledge and understanding of content through developed and accurate descriptions, explanations, and examples.

Assessment

Students will grade the activity (*Handout 4.2G*).

Cognitive Reflection

How did doctors justify using patients in public hospital wards as medical research subjects without obtaining their consent or offering them financial compensation (RI. 6.7, RI 7.7, RI.8.7, RI. 9–10.7)? Do you agree with their reasoning? Explain your answer.

Keep, Retain, and Generalize

Who did what with the cells, when, where, and for what purpose? Who benefited, scientifically, medically, and monetarily (RST.9–10.6, RST.11–12.6)?

LESSON 4.3. SCIENCE AND CLEAN WATER

The Boy Who Harnessed the Wind by William Kamkwamba and Bryan Mealer (2010)

Summary. The novel *The Boy Who Harnessed the Wind* is about the moving story of a boy who lives in a small African village in Kasungu, Malawi. As a young student, William is fascinated with electronics and often spends his free time searching his local junk yard for parts to refurbish and build electronics. His interest in engineering leads him to become the unwitting savior of his village when he is able to build a windmill to access clean water to save the crops of his village in a time of extreme drought and famine.

Application

Goal. Plan and conduct an investigation of the properties of water and its effects on Earth's materials and surface processes.

Objectives. Students will be able to do the following:

- Investigate the importance of properties of water and its impact on surface processes in food production in third-world countries.

- Analyze nonfiction texts, including how an author's ideas or claims are developed and how accounts of a subject are told in different mediums.

Standards. To view the Common Core Standards that correspond with this lesson, please visit the *STEAM Meets Story* page on www.tcpress.com and click on the Resources tab.

Teaching Strategy

Introduction. In the movie *Black Panther* (Coogler, 2018), based on the Marvel comic series, the main hero, T'Challa, resides in the technologically advanced country of Wakanda. Wakanda is known as one of the richest countries in the Marvel comic universe. Compare the fictional country of Wakanda and its successes to the modern-day African country of Malawi.

Discussion. (RI.9–10.5, RI.9–10.6, RL. 6.7, RL.7.7, RL.8.7, RL.9–10.7)

- How does access to clean potable water sources impact many African countries' ability to grow economically and technologically like Wakanda?

- Students will reflect on the various ways they use water in their own communities to activate prior knowledge and student interest.

Materials

- Handouts
- Selected video clips
- Computers
- Copy of excerpts from the book, *The Boy Who Harnessed the Wind*

Time. One to two class periods

Essential vocabulary. Using a list of words for students related to science from the book *The Boy Who Harnessed the Wind*, students will create a foldable graphic organizer for vocabulary and a model of the water cycle to help retain important details (WHST.9–10.2.9). A sample listing of key vocabulary words that will help the delivery of the lesson is shown in *Handout 4.3A*.

Handout 4.3A. Key Vocabulary for Science, English, and Language Arts

Word	Definition	Sentence
Condensation		
Evaporation		
Precipitation		
Transpiration		
Water pollution		
Aquifer		
Saturated zone		
Unsaturated zone		
Wetlands		
Runoff		
Collection		
Infiltration		
Absorption		
Excretion		
Recharge		
Groundwater		
Surface water		
Estuary		
Author's point of view		
Biography		
Connotative		
Denotative		

Handout 4.3B. Science Investigation

My question:

My prediction:

How will you do this?

1.	2.	3.

_____ _____ _____

_____ _____ _____

_____ _____ _____

FAIR TEST

I will keep the same	I will change

Trial/Try-Out

Students will watch a brief video on the water cycle and water sources (https://www.khanacademy.org/science /high-school-biology/hs-ecology/hs-biogeochemical -cycles/v/the-water-cycle). Students will create a hypothesis about the potability of water from different environments (e.g., swamp, city, ocean, well) and test water samples for contamination from different sources.

Students will answer questions on the implications of contaminants they may find in their different water supplies.

Assessment

Based on their results, students will create a problem-solving project that addresses how contaminants found in the water sources impact human survival and economic growth.

Like William Kamkwamba in *The Boy Who Harnessed the Wind*, students will be required to select a city or country without access to potable water and provide an analysis of economic problems that contribute to access to clean water. Students will also be required to develop a proposal that addresses how these countries can use other sources of clean energy (solar, wind) to improve access to clean water (RI. 6.7, RI.7.7, RI.8.7, RI.9–10.7). *Handout 4.3B* addresses the scientific investigation process.

Cognitive Reflection

Students will watch a news clip on the Flint, Michigan, water crisis and reflect in their journals on how this city was disrupted by a lack of access to clean water as well as the causes of the crisis (RI. 6.7, RI.7.7, RI.8.7, RI.9–10.7).

Keep, Retain, and Generalize

Students will analyze William Kamkwamba's point of view on his ability to help his village in the book *The Boy Who Harnessed the Wind* and how his actions can impact their own attempts to help solve the water issues experienced in their project (RI. 6.7, RI.7.7, RI.8.7, RI.9–10.7).

LESSON 4.4. SCIENCE, MEDICINE, AND DRUGS

Percy Jackson and the Lightning Thief by Rick Riordan (2005)

Summary. In *Percy Jackson and the Lightning Thief*, Percy finds himself at Camp Half-Blood, a training camp for demigods like him. He discovers that he is a demigod son of Poseidon, the Greek god of the sea; his best friend Grover is a satyr that the gods are accusing of having stolen Zeus's master lightning bolt. To clear his name and save the world from another war between the Olympian gods, he sets out to retrieve the lightning bolt from Hades, who they believe is the real thief. Thus, Percy Jackson and his companions, Grover

Underwood and Annabeth Chase, a daughter of Athena, start on a journey to the underworld, facing numerous mythological monsters on the way.

Application

Goals. Students will understand the following:

1. The different kinds of mythology
2. Different uses for drugs

Objectives. Students will do the following:

- Name three positive and three negative uses for drugs.
- Identify at least five different myths from Greek, Roman, and Egyptian methodology.

Standards. To view the Common Core Standards that correspond with this lesson, please visit the *STEAM Meets Story* page on www.tcpress.com and click on the Resources tab.

Teaching Strategy

Introduction. Students will watch a clip of Percy and his friends as they search for the lightning thief (the clip is taken from the film adaptation of the book). On their search, they go to a casino where a flowering lotus plant (opiate) is given to them (https://www .youtube.com/watch?v=3yV0I3lKito) (RI. 9–10.7, RST. 9–10.6, L.6–8.6, L. 9–10.6, L9-10.4).

1. What kinds of things happened to them and others who continually ate the plant?
2. Can opioids do this to you today?
3. What is the chemical makeup of an opioid? How are opioids helpful? How are they dangerous?

4. What area of STEM would you become interested in if you developed opioids?

Materials

- Handouts
- Selected video clips
- Computers
- Copy or excerpts from the books *Percy Jackson and the Lightning Thief*

Time. One to two class periods

Essential vocabulary. The list of words for students related to science from the book *Percy Jackson and the Lightning Thief* are listed in *Handout 4.4A* (RI.9–10.7, RI.8.7, RI.7.7,RI. 6.7).

Handout 4.4A. Essential Vocabulary

Directions: What are some of the science-related words from *Percy and the Lightning Thief* that are similar to the current unit in science in which you are presently working?

Words from Percy Jackson	Words from Your Science Unit or Book or Drug and Medicine vocabulary	Definitions
	Addicted	
	Addictive	
	Chemical	
	Codeine	
	Fentanyl	
	Heroin	
	Hydrocodone	
	Illegal	
	Medication	
	Methadone	
	Misuse	
	Morphine	

Handout 4.4B. What Knowledge Do You Need to Know About Drugs

Directions: Address the differences in the information these three groups would need to know about drugs.

Parents	Teacher	Students

Handout 4.4C. Positive and Negative Effects of Drugs

Directions: Name three positive and three negative effects of drugs. Elaborate on each.

Positive Effects	Negative Effects
1.	1.
2.	2.
3.	3.

Trial/Try-Out

Activity 1

Have students browse the article "What You Need to Know About Drugs" (https://kidshealth.org/en/kids/know-drugs.html). On the handout, determine the difference in what parents, students, and teachers need to know. (*Handout 4.4B*) (RST.6–8.5, RST. 9–10.5. RST.6–8.4, RST. 9–10.4). Allow students to address the following questions in pairs or groups:

- How does the information differ according to the category?
- What kinds of things should you be aware of related to drugs?

Activity 2

Drugs have many positive purposes. Place students in a group and have them decide the following:

- How are drugs helpful?
- What persons/careers might be involved (*Handout 4.4C*) (RI.6.7, RI.7.7, RI.8.7, RI. 9–10.7, L.6.4.c, L.7.4.c, L.8.4.c, L.9–10.4.c)?

Assessment

Group work. Start a campaign at school or in the community to address improper drug use (RI.6.7, RI.7.7, RI.8.7, RI. 9–10.7) (*Handout 4.4D*).

a. Develop a flyer. Decide what pertinent information is needed.
b. Make a commercial for TV. What needs to be in the script? How can you make it eye-catching and convincing? What facts are most important?

Handout 4.4D. Design a Flyer

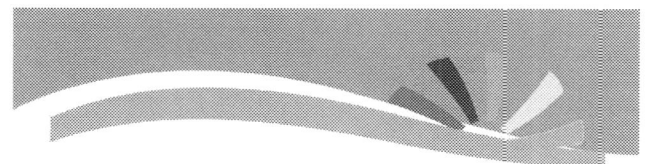

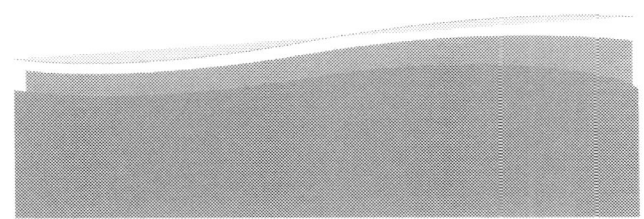

c. Design a presentation. What data do you need for a presentation?
d. You want to convince someone to become a person who provides drugs for positive purposes. What would you say to urge them to adopt that as a career?

Cognitive Reflection

Have students search the Internet in teams to answer the following:

1. What are topics in the news that speak to the helpfulness of drugs?
2. What topics in the news speak to the destruction that drugs can cause?
3. What drugs are usually abused (RI.6.7, RI.7.7, RI.8.7, RI. 9–10.7) (*Handout 4.4E*)?
4. What can you do as a student to have a positive impact on the drug culture? Is drug addiction a worldwide problem? What happens in other countries with drugs?

Keep, Retain, and Generalize. (RL.6.7, RL.7.7, RL.8.7, RL. 9–10.7)

a. Have students read selected chapters or excerpts from *Percy Jackson and the Lightning Thief*.
b. Students will perform a think-pair-share and brainstorm with one or two students in the class. They will note different pictures:
  • A Greek mythology book
  • A Roman mythology book
  • An Egyptian mythology book
c. Have the students view a film about Egyptian mythology (https://www.youtube.com/watch?v=uZe49S1Q8b8). From the pictures and film provided, address the following questions:
  • Why are the stories in Greek, Roman, and Egyptian mythology similar?
  • What culture created the stories first?
  • Design a collage to demonstrate what you have learned (*Handout 4.4F*).

Handout 4.4E. Read All About It

Helpfulness of Drugs	Destructive Drugs	Drugs Usually Abused
1.	1.	1.
2.	2.	2.
3.	3.	3.
4.	4.	4.

Handout 4.4F. Design a Collage

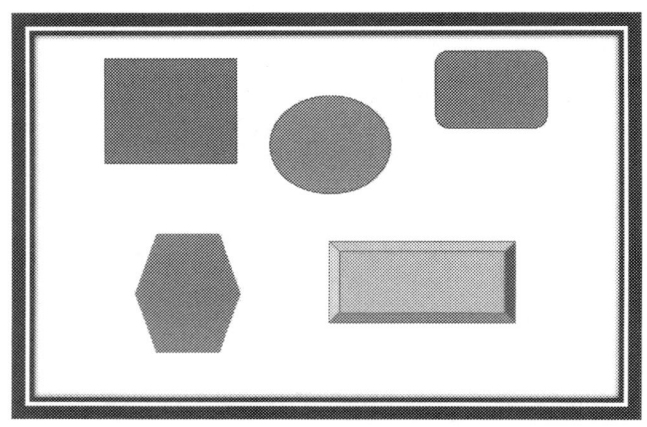

LESSON 4.5. GRAVITY, WEIGHT, AND MASS

Ender's Game by Orson Scott Card (2014)

Summary. The novel tells the story of a young boy, Ender Wiggin, who is sent to Battle School Training Academy, located in orbit above the Earth, built to train people to become soldiers who will one day battle against a vast alien race known as Buggers. Ender tries his best to get promoted in the difficult training scheme; his brother and sister are trying to restore the world and to make it a better place. For Ender, the training is tough, but he excels over his classmates and becomes a leader. He is granted a very special teacher, who helps him to become a commander to save humanity from an invasion from another world.

Application

Goal. Students will learn to support an argument that the gravitational force exerted by Earth on objects is directed down.

Objectives. Students are engaged in the following scientific practices and will do the following:

- Calculate their weight on other planets by using an equation.
- Use their data to describe how gravity affects weight.
- Answer what gravity has to do with weight and mass.
- Make observations of various objects falling to the ground to provide evidence of the effects of gravity from Earth.

Standards. To view the Common Core Standards that correspond with this lesson, please visit the *STEAM Meets Story* page on www.tcpress.com and click on the Resources tab.

Teaching Strategy

Introduction

1. There is a difference between mass and weight as gravity acts upon one but not the other. Based on this understanding, the students will calculate their weight on several planets where the force of gravity is not the same as Earth's. Students will then construct an explanation about the gravitational force on an object (RST.9–10.6, RI.9–10.7, L.6.6, L.7.6, L.8.6, L.9–10.6)?
2. Students will watch several clips from the film adaptation of *Ender's Game*. Ender is in zero gravity and is being trained to be a leader and a warrior. Ender has two main scenes in the movie where he talks about gravity (RL.6.7, RL.6.7, RL.6.8.7, RL.6.9–10.7, RL.6.11–12.7).
   - He examines the war game zero gravity in an artificial space as he leads his team (https://www.youtube.com/watch?v=XKy8-QiVWm4).
   - He practices his strategies. He uses his knowledge of gravitational forces to win against the opposing team during war games (https://www.youtube.com/watch?v=2NPSwpdx6iQ).
   a. Students will read the article "The Difference Between Weight and Mass" (https://cosmosmagazine.com/physics/explainer-what-s-the-difference-between-mass-and-weight/) to study the formula regarding the difference between weight, mass, and acceleration. Weight is a measure of the force of gravity pulling down on an object. It depends on the object's mass, which is how much matter the object contains. It also depends on the downward acceleration of the object due to

gravity, which is the same all over Earth. Weight can be represented by the equation $F = m \times a$. This is the general equation that relates force to mass and acceleration. When it relates weight to mass and acceleration, the letter F represents the object's weight in Newtons (N), which is the SI unit for weight. The letter m in the equation represents the object's mass in kilograms, and the letter a represents the downward acceleration due to gravity. As this equation shows, weight is directly related to mass. As an object's mass increases, so does its weight. For example, if mass doubles, weight doubles as well. Students will solve the equation (RI.6.7, RI.7.7, RI.8.7, RI.9–10.7).

Materials

- Handouts
- Selected video clips
- Computers
- Copy or excerpts from the book

Time. One to two class periods

Essential vocabulary. The words related to science from the book *Ender's Game* are listed in *Handout 4.5A*.

Trial/Try-Out

1. How does gravity affect weight and mass of an object? Students will explain the difference between weight and mass.

Handout 4.5A. Essential Vocabulary

Key Terms in Your Text Related to Weight and Mass	Definition
Air resistance	
Balanced forces	
Force	
Friction	
Gravity	
Net force	
Unbalanced forces	
Weight	
Inertia	
Newton's first law of motion	

2. Students will determine the lightest to the heaviest a person would be on the surface of each planet in our solar system. On which planet in our solar system do you think an object or person would be the heaviest? On which planet would they be the lightest? Calculate their weight on another planet using multiplication [mass (on Earth) × gravity = weight] (RST.9–10.6).

a. Students in groups of four will determine how far they can jump on Earth. Then they will calculate their jump length on other planets and the moon using this equation: Earth jump length average divided by the planet or moon's gravity.

Assessment

Why do you think that the planets differ? Include details about gravitation force on an object (L.6.6, L.7.6, L.8.6, L.9–10.6).

Cognitive Reflection

1. Read excerpts from the ending of *Ender's Game* and/or view the cited clip (https://www.youtube.com/watch?v=6RVyL8lNtj4) (WHST 6–8.2.d, WHST. 9–10.2.d, RL.6.7, RL.7.7, RL.8.7, RL.9–10.7).
2. Ender and his team believe they are fighting in a simulation. When he finds that this is not simulation, he realizes he has destroyed a complete species. Although human ethnic groups and races do not constitute distinct species, there are historical instances wherein cultures and peoples have been almost destroyed by genocide. Prominent examples from modern history include the killing and displacement of Native Americans by European settlers of North America and the systematic targeting and killing of Jews during the Holocaust (WHST 6–8.2.d, WHST. 9–10.2.d, RL.6.7, RL.7.7, RL.8.7, RL.9–10.7).

- Jewish people (https://www.ajhs.org/essential-readings)
- Native Americans (https://www.loc.gov/teachers/classroommaterials/presentationsandactivities/presentations/immigration/native_american.html)
- In teams of four, decide which group to read about (Jewish people or Native Americans).
  » What happened to the Native Americans or Jewish people?
  » Why did it happen?
  » Who were the antagonists?
  » Was it justified?

Keep, Retain, and Generalize

List other cultures that have experienced genocide. What similarities can you draw between the experiences of Native Americans and Jewish people and the experiences of other victims of genocide? (WHST 6–8.2.d, WHST. 9–10.2.d, RL.6.7, RL.7.7, RL.8.7, RL.9–10.7).

LESSON 4.6. RESPIRATORY SYSTEM

After Earth by Peter David (2013)

Summary. Since humanity's exodus from the Earth a thousand years ago and the attack of an alien force that followed, Cypher Raiges was a ranger who fought the Ursas. Cypher takes his teenage son, 13-year-old Kitai, on a trip to another planet and they crash on Earth when an asteroid collides with their craft. Cypher is seriously wounded and Kitai must venture out into now-hostile territory: Earth. The Ursas, large predatory creatures that hunt by "sensing" fear, were released in the crash. Cypher learned how to completely suppress his fear, a technique called "ghosting," so that he could kill the aliens. Cypher instructs Kitai to locate the distress beacon in the tail section of the ship, which broke off while descending to Earth. Cypher gives Kitai his weapon, a wrist communicator, and six capsules of a fluid that enhances the oxygen intake so he can breathe in Earth's low-oxygen atmosphere. Cypher warns him to avoid the highly evolved wildlife and vegetation and to be careful of violent thermal shifts. Kitai leaves to find the tail section, with Cypher guiding him through the communicator.

Application

Goals. Students will learn about the following:

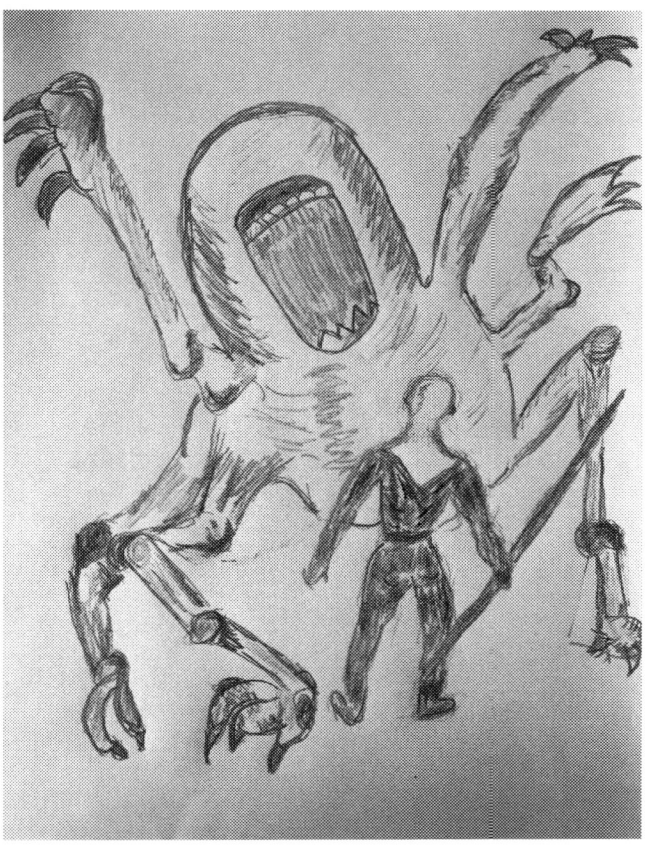

- The respiratory system
- Pheromones
- Going past fear to move into your career
- Environmental challenges

Objectives

- Students will be engaged in scientific practices. Students will be able to identify the following:
  - » The parts of the respiratory system and what negatively affects it
  - » The purpose of pheromones and their effect on living things
  - » Ways to address fear to seek the career you want.
- Students will be able to name at least five reasons why plants, animals, and insects adapt.

Standards. To view the Common Core Standards that correspond with this lesson, please visit the *STEAM Meets Story* page on www.tcpress.com and click on the Resources tab.

Teaching Strategy

Introduction

In a group of four, discuss the following: What does the respiratory system do? What are the parts of the respiratory system? What are things in the air that can affect the respiratory system?

1. COVID-19 affects the respiratory system in what kinds of ways?
2. Draw a diagram with your group and be ready to discuss and connect the varied parts of the respiratory system where COVID has affected the system (RST. 9–10.6).

Materials

- Handouts
- Selected video clips
- Computers
- Copy or excerpts from the book

Time. One to two class periods

Essential vocabulary. Words related to science from the book *After Earth* are listed in *Handout 4.6A*.

Trial/Try-Out

Kitai and Cypher land on Earth and every system on Earth is negatively affected and has become hostile toward humans, including plants, insects, and animals. Kitai encounters some primates (https://www.youtube.com/watch?v=t-lIuwPGT9w). In another scene (https://www.youtube.com/watch?v=eMgfTq1Z2n8) Kitai is affected by an insect bite (WHST.6–8.2, L.6.6, L 7.6, L.8.6, L.9–10.6).

1. What does an environmental scientist do?
2. If you were an environmental scientist, what would be your theory as to why Earth revolted against humankind?
3. What kind of environmental issues do we have going on today?
4. In a team of four, use your research tools to find out the following:
   a. What is happening to the air?
   b. How does that affect the respiratory system?
   c. What solutions can you research and find?

Handout 4.6A. Essential Vocabulary

Directions: What are some of the science-related words from *After Earth* that are similar to the current unit in science in which you are presently working?

Words from *After Earth*	Words from Your Science Unit or Book or Respiratory Vocabulary	Definitions
	Respiration	
	Artery	
	Breathing	
	Bronchus	
	Capillary	
	Cartilage	
	Cellular Respiration	
	Circulation	
	Combustion	
	Exchange	
	Exhale	
	Expel air	
	Inhale	
	Rib cage	
	Trachea	
	Transport	
	Vein	
	Voice box	

Assessment

Devise a clean air campaign. Make a pamphlet or a presentation detailing how your respiratory system is affected by air pollution (RI.6.7, RI.7.7, RI.8.7, RI.9–10.7, RST.9–10).

Cognitive Reflection

In the film clips, Kitai has to take a chemical that helps him breath on planet Earth. There are people who are ill who cannot breathe properly. What kinds of medicine do they take (RI.6.7, RI.7.7, RI.8.7, RI.9–10.7, RST.9–10)?

Keep, Retain, and Generalize

Cypher, Kitai's dad, knows how to ghost, a procedure where he learns to not show any fear. The Ursa were sent to Nova Prime where Cypher works as a ranger. The rangers protect the people. The Ursa cannot see Cypher because he has learned to control his fear. In one of the final scenes of the movie adaptation of *After Earth*, Kitai learns how to ghost (https://www.youtube.com/watch?v=0vcXvLUt1E4) (RI.6.7, RI.7.7, RI.8.7, RI.9–10.7, RST.9–10).

1. What kind of pheromones do humans project? How do our body systems change when we are afraid? What does "the smell of fear" mean?
2. What kind of fear might you experience when deciding to train for a STEM career? What can you do to decide if you would like a STEM career or if you have the aptitude for it? Why are there fewer women in STEM careers? (*Handout 4.6B*)

Handout 4.6B. Career Exploration

What STEM career would you like to pursue? What are the steps you would take to move toward this goal?

- Step 1

- Step 2

- Step 3

LESSON 4.7. GENETICS

Gattaca, Written and Directed by Andrew Niccol (1997)

Summary. The story revolves around a person who defies a genetics-obsessed system of perfection, with perfection now controlling existence on a future Earth. Vincent is an "In-Valid" who assumes the identity of a member of the genetic elite to pursue his goal of traveling into space with the Gattaca Aerospace Corporation, which is about to launch to the moon. The week before his mission, a murder marks Vincent as a suspect. A colleague with whom he has been dating begins to realize Vincent is not who he pretends to be as he is pursued by a relentless investigator and Vincent's dreams steadily unravel.

Application

Goal. Understand the concepts of genetics.

Objectives. Students will be able to do the following:

- Trace a gene as it is passed down from generation to generation.
- Apply concepts of statistics and probability to explain the variation and distribution of expressed traits in a population.
- Complete a Punnett square for dominant and recessive traits.
- Learn vocabulary words related to genetics, such as homozygous, heterozygous, dominant, recessive, genotype and phenotype.

Standards. To view the Common Core Standards that correspond with this lesson, please visit the *STEAM Meets Story* page on www.tcpress.com and click on the Resources tab.

Teaching Strategy

Introduction. Vincent Freeman, a male in the future, is conceived without the aid of genetic selection. His genetic profile indicates a high probability of several illnesses and a projected life span of around 30 years. His parents used genetic selection in conceiving his brother.

Sickle-cell anemia is an autosomal recessive genetic disorder that causes red blood cells to change shape, which can cause the red blood cells to become stuck in blood vessels. Write in the genotypes on the line next to/below each individual (RST.9–10.6, RST. 11–12.6, RI.6.7, RI.7.7, RI.8.7, RI.9–10.7).

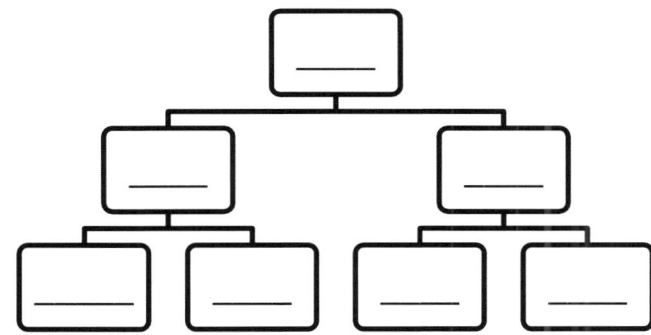

Sickle cell is a trait to which African Americans are predisposed. Tay-Sachs disease is a rare, inherited neurodegenerative disease. People with Tay-Sachs disease do not have enough of an enzyme called beta-hexosaminidase A. The less enzyme a person has, the more severe the disease and the earlier that symptoms appear. There are three forms of Tay-Sachs disease, distinguished by the general age of onset. What are some other rare genetic disorders? Are any of these disorders more common among particular populations? Investigate and find what some other groups may be predisposed to (RI.6.7, RI.7.7, RI.8.7, RI.9–10.7).

Materials

- Handouts
- Selected video clips
- Computers

Time. One to two class periods

Essential vocabulary. Words related to science from the textbook and a movie on genetics are provided in *Handout 4.7A* (L.6.6, 7.7.6, L.9–10.6).

Trial/Try-Out

In the future world of *Gattaca*, eugenics is common. Genetic discrimination is illegal, but in practice genotype profiling is used to identify "valids" for professional employment while "in-valids" are given menial jobs (WHST.9–10.2.d, RL 8.7, RL.9–10.7).

Handout 4.7A

What words did you find that are related to genetics? What are the definitions of those words?

Words from the Movie	Words from the Text	Definitions

In July 2013 a worldwide protest movement, Black Lives Matter, was introduced. It received widespread recognition in 2020. African Americans have been thought to have inferior genes, an assumption that has contributed to the unjust conditions that movements like Black Lives Matter are protesting. Worldwide, what other groups of people were thought to be inferior (e.g., French Canadians, Irish immigrants, Jews, Native Americans)?

- Why are there caste systems?
- Get in groups of four. Discuss whether some groups of people are really inferior.

Assessment

Develop a dialogue for a play to demonstrate the pros and cons of being supposedly inferior (RI.6.7, RI.7.7, RI.8.7, RI.9–10.7).

Cognitive Reflection

Students can choose one of the following topics for discussion:

1. What happens when people from different races have children? Do they get the "best" of both worlds? Research with your group and debate this concept.
2. Is genetic engineering close to cloning?
3. How are DNA and genetics related?

Keep, Retain, and Generalize

In a video clip (https://www.youtube.com/watch?v=xsnqsAzCoUY), Vincent has to prove to Anton that he belongs at Gattaca and he is just as good as those who are genetically engineered (RL.6.7, RL.7.7, RL.8.7, RL.9–10.7).

Do certain groups have to prove their worth? What persons from nondominant groups have famous STEM careers?

LESSON 4.8. ENERGY

The Matrix, Written and Directed by the Wachowskis (1999)

Summary. Computer programmer Thomas Anderson, known as Neo, has a strange feeling about the world and is puzzled by repeated online encounters with the phrase "the Matrix." Trinity contacts him and tells him a man named Morpheus has the answers he seeks. Neo meets Morpheus, who offers him a choice between two pills: red to reveal the truth about the Matrix and blue to return him to his former life. After Neo swallows the red pill, his reality falls apart, and he awakens in a liquid-filled pod. Morpheus explains the truth. In the early 21st century, there was a war between humans and intelligent machines. When humans blocked the machines' access to solar energy, the machines harvested the humans' bioelectric power, keeping them pacified in the Matrix, a shared simulated reality of the 20th century.

Application

Goal. Students will learn about alternative fuel and energy sources.

Objectives. Students will know at least five sources of energy and will do the following:

- Describe sources and uses of energy.
- Define renewable and nonrenewable energy.
- Provide examples of common types of renewable and nonrenewable resources.
- Understand and explain general ways to save energy at a personal, community, and global level.
- Understand and explain, in general terms, how passive solar heating, hydropower, and wind power work.
- Describe some general characteristics of solar power, hydropower, and wind power.
- Understand the benefits and disadvantages to using renewable resources.

Standards. To view the Common Core Standards that correspond with this lesson, please visit the *STEAM Meets Story* page on www.tcpress.com and click on the Resources tab.

Teaching Strategy

Introduction. In *The Matrix*, humans were an energy source (https://www.youtube.com/watch?v=Iojq OMWTgv8).

Americans experience high prices at the gas pumps, and winter heating and summer coolant are costly. Consumers are looking for ways to lower their energy costs. Our primary energy sources today are fossil fuels, which are almost exhausted. Scientists have long warned of the environmental damage caused by burning fossil fuels. Americans must use and develop alternate forms of energy without destroying the Earth's environment. Exploring the use of renewable and alternative resources is a must. There are various types of renewable energy resources (solar, water, and wind) that can be transferred into electricity. Energy can move and change, but it cannot be destroyed. Almost every form of energy can be converted into other forms. Remember you can hear energy (sound), feel energy (wind), taste energy (food), and see energy (light). Energy originally comes from the sun. Other energy is called nonrenewable because once it is used up it is gone, like coal and oil. In pairs, brainstorm a list of ideas about where and when we use energy (WHST.6–8.2.9, RI.9–10.7, RST9-10.6, RST.11–12.6).

Materials

- Handouts
- Selected video clips

Handout 4.8A. Vocabulary

Biomass is the combustion of materials that originate from living things.

Chemicals are used to fuel automobiles and other vehicles.

Electrical fuel drives many small machines and keeps lights glowing.

Geothermal energy taps steam from water heated underground (like geysers) and uses it to spin turbines.

Hydrogen power uses electricity to break down water into hydrogen gas. The amount of energy released is less than the energy used to break it apart, so it is not currently feasible.

Hydroelectricity generates electricity by harnessing the power of flowing water (a renewable resource as long as there is rain).

Kinetic energy is the energy of motion. A spinning top, a falling object, and a rolling ball all have kinetic energy. The motion, if resisted by a force, does work. Wind and water both have kinetic energy.

Light energy is generated from light bulbs and computer screens and the sun.

Nuclear fusion imitates the method the sun uses to produce energy. It involves the joining together of the nuclei of hydrogen atoms.

Nuclear fission is when energy is given off from splitting nuclei of uranium atoms.

Potential energy is the energy stored by an object as a result of its position (e.g., roller coaster at the top of a hill).

Sound energy is created (e.g., when a door slams, it releases sound energy).

Solar energy occurs from the sun (light).

Thermal energy (or heat) boils water, keeps us warm, and drives engines.

Tidal energy is when the energy from ocean tides is harnessed.

- Computers
- Copy or excerpts from the book

Time. One to two class periods

Essential vocabulary. Words that are various forms of energy are listed in *Handout 4.8A*.

Trial/Try-Out

Imagine yourself having surgery in a hospital and the power goes out. Generators are like storage houses for energy and are usually powered by electricity from coal or fossil fuels. It is better if everything is backed up with renewable power like solar energy or another type of stored renewable energy. The renewable source is always supplying more energy (i.e., the sun is almost always shining, wind is always blowing, and rivers are always running). Storing renewable energy for power failures is a better idea because those energy supplies will never run out. There are several types of energy sources listed in *Handout 4.8A* (L.6–8.4, L.9–10.4).

1. In groups, students will research how solar energy is used to heat buildings by investigating the thermal storage properties of some common materials: sand, salt, water, and shredded paper.
2. Then students will evaluate the usefulness of each material as thermal storage and how the sun can be used for heating.

Assessment

In groups, students will do the following (RST.11–12.6, RST. 9–10.6):

1. Research how the wind is used to generate electricity.
2. Build a model anemometer to better understand and measure wind speed (https://www.youtube.com/watch?v=Af0LB3abBsk).
3. Observe a model of a working waterwheel to investigate the transformations of energy involved in turning the blades of a hydro turbine (https://www.youtube.com/watch?v=CiEUvJR1zUc).
4. Write and explain the characteristics of hydropower plants (*Handout 4.8B*).

Cognitive Reflection

1. How would you feel if you were no longer able to go on school field trips or participate in extracurricular activities because of the high cost of transportation to and from these events?
2. How do you think your learning would be affected if you were in a classroom that was only heated to 60 or 65 degrees (RST.11–12.6, RST. 9–10.6)?

Keep, Retain, and Generalize

1. The average cost of a gallon of gasoline was $2.10 one year ago. Today, the average cost of a gallon of gasoline is $3.00. Calculate how much more it is

Handout 4.8B

Write and explain the characteristics of hydropower plants.

Characteristics

1.

2.

3.

4.

5.

6.

costing you to fill your car with gas each month, assuming that you fill your car once each week and there are 4 weeks in each month (RST.11–12.6, RST. 9–10.6).
2. Home heating bills typically rise during the winter because of the colder weather. Calculate how much this increase will amount to each month if your typical home heating bill was $150 per month last winter (RST.11–12.6, RST. 9–10.6).

LESSON 4.9. DESERT PLANTS AND ANIMALS

Holes, by Louis Sachar (2008)

Summary. Stanley has been unjustly sent to a boys' detention center, Camp Green Lake, where the warden makes the boys "build character" by spending all day, every day, digging holes 5 feet wide and 5 feet deep. It doesn't take long for Stanley to realize there's more than character improvement going on at Camp Green Lake. The boys are digging holes because the warden is looking for treasure. Stanley tries to dig up the truth in this inventive and darkly humorous tale of crime and punishment—and redemption.

Application

Goal. Students will understand plants and water supply in the desert.

Objectives. Students will be able to identify the following:

- Plant vegetation in the desert
- Water supply and droughts in the desert

Standards. To view the Common Core Standards that correspond with this lesson, please visit the *STEAM Meets Story* page on www.tcpress.com and click on the Resources tab.

Teaching Strategy

Introduction. Stanley and Zero escape from barren Camp Green Lake and take refuge atop God's Thumb, filled with vegetation and an abundant water supply. Stanley and Zero are covered with the deadly, venomous, yellow-spotted lizard but are not bitten due to their consumption of onions. Additionally, flashbacks in the movie reveal the medicinal uses of the onion (e.g., hair growth). The onions acted as a repellent and the lizards would not bite (https://www.youtube.com /watch?v=Ee61QYQxG5o).

Students will read about desertification (https:// earthobservatory.nasa.gov/features/Desertification /desertification2.php) and then answer the following questions:

1. What causes desertification?
2. What type of data should be collected to study desertification?
3. Using the data provided identify periods of drought.

4. Provide a comparative analysis of vegetation and rainfall according to the line graph provided.

Materials

- Handouts
- Selected video clips
- Computers
- Copy or excerpts from the book *Holes*

Time. One to two class periods

Essential vocabulary. Words related to science from the book and movie are provided in *Handout 4.9A* (L.6.6, 7.7.6, L.9–10.6).

Trial/Try-Out

1. Deserts are found in dry continental interiors away from the coasts. They are found in a belt at approximately 30 degrees north and south of the Equator. Temperatures are hot all year round as the sun is overhead. The daily maximum temperature is over 40 degrees. It can fall to below freezing at night due to the lack of cloud cover. Vegetation is sparse and usually confined to water.
   a. What are some of the characteristics of desert plants such as the cacti and other succulents? What about leafy plants?

Handout 4.9A. Essential Vocabulary

Directions: What are some of the science-related words from *Holes* that are relevant to the current unit in science in which you are presently working?

Words from *Holes*	Words from Your Science Unit or Book	Definitions

b. What happens to seeds and seed germination in the desert?

c. What is the average rainfall in the desert?

2. Desert rodents such as the hamster are small and nocturnal and live largely underground. Being small reduces the amount of food required. By being active at night and staying underground, they avoid the heat of the day and reduce their need for water. What other animals are in the desert?

3. What insects are in the dessert? What are their characteristics?

4. How much rainfall does the desert receive in 1 year?

Assessment

The people of the desert are often nomadic and move from place to place using camels for transport. How do these people live (RST.11–12.6, RST.9–10.6, RI.9–10.7)?

Watch the film (https://www.youtube.com/watch?v=QbzFiWNhFGI) and describe what desert soils are like.

Cognitive Reflection

Stanley's dad has spent his life attempting to find a cure for smelly shoes (caused by *Brevibacterium*). He conducted countless odor-eliminating experiments. He finally cures smelly shoes with Sploosh, a combination of onions and peaches. What other medicines have been derived from the desert (RST.11–12.6, RST. 11–12.6, RST.9–10.6, RI.9–10.7)?

Keep, Retain, and Generalize

Why does the Earth need deserts in the ecological system (RST.11–12.6, RST.9–10.6, RI.9–10.7)? Use *Handout 4.9B* to list living things found in deserts and their characteristics.

Handout 4.9B. Characteristics

What are some desert plants, animals, and insects? What are their characteristics?

	Plants	Animals	Insects
1.			
2.			
3.			
4.			
5.			
6.			

LESSON 4.10. ASSISTIVE TECHNOLOGY

Star Wars: Splinter of the Mind's Eye, by Alan Dean Foster (1978)

Summary. The plot of the story involves Luke Skywalker and Princess Leia, who are marooned together on a world far away, where they encounter struggles against the forces of the evil Galactic Empire and Darth Vader.

Application

Goal. Students will learn about the science of robotic prosthetics.

Objective. Students will examine the neuroscience of prosthetics as a career.

Standards. To view the Common Core Standards that correspond with this lesson, please visit the *STEAM Meets Story* page on www.tcpress.com and click on the Resources tab.

Teaching Strategy

Introduction. Darth Vader was controlled by robotics. Luke Skywalker had a robotic arm. The University of Chicago Medical Center's neuroscientists have shown how amputees can learn to control a robotic arm through electrodes implanted in the brain. The research details changes that take place in both sides

of the brain used to control the amputated limb and the remaining, intact limb. The results show both areas can create new connections to learn how to control the device, even several years after an amputation. Currently, more than 1 million annual limb amputations are carried out globally due to accidents, war casualties, cardiovascular disease, tumors, or congenital anomalies. Study in robotic prosthetic limbs is a well-established research area that integrates advanced mechatronics, intelligent sensing, and control for achieving higher order lost sensorimotor functions while maintaining the physical appearance of an amputated limb. Robotic prosthetic limbs are expected to replace the missing limbs of an amputee, restoring lost function and providing an aesthetic appearance. The main aspects for the amputee are enhanced social interaction, a more comfortable life, and productive contribution to society. With the advancement of sensor technology, in the last few decades significant contributions have been made in this area. Much of the work is still in the research stage, and more research and development work is anticipated in the coming years, with the ultimate goal to produce a device that can generate human-like motions.

Goal. Students will understand the need and use of enhanced prosthesis and be aware of scientific careers.

Objective. Students will be able to name and provide the usage for enhanced technology for adults and persons with disabilities.

Materials

- Handouts
- Selected video clips
- Computers
- Copy of or excerpts from books from the *Star Wars* series

Time. One to two class periods

Essential vocabulary. Words related to science are provided in *Handout 4.10A* (L.6.6, 7.7.6, L.9–10.6).

Trial/Try-Out

How can these types of technologies help persons with disabilities? Technology can level the playing field for students with mobility, hearing, or vision impairments. Technology has opened many educational doors to children, particularly to children with disabilities. Alternative solutions from the world of technology are accommodating physical, sensory, or cognitive impairments in many ways. Much of the technology we see daily was developed initially to assist persons with disabilities. Curb cuts at street corners and curb slopes, originally designed to accommodate people with orthopedic disabilities, are used more frequently

Handout 4.10A. Essential Vocabulary

Directions: What are some of the science-related words from *Star Wars* that are similar to the current unit in science in which you are presently working?

Words from *Star Wars*	Words from Your Science Unit or Book	Definitions

Handout 4.10B

How does assistive technology differ for these three populations?

The Elderly	Students With Disabilities	General Population

by families with strollers or individuals with grocery carts than by persons with wheelchairs or walkers. The optical character reader, developed to assist individuals unable to read written text, has been adapted in the workplace to scan printed documents into computer-based editable material, saving enormous amounts of data entry labor (RST.11–12.6, RST. 11–12.6, RST.9–10.6, RI.9–10.7) *Handout 4.10A* and *Handout 4.10B*.

In a group of four, students will list the following:

1. Various types and categories of assistive technology for persons with disabilities
2. Various types of categories and types of assistive technology for the elderly
3. Various types of categories and types of assistive technology for the general population

Assessment

How are the technologies for the elderly, students with disabilities, and the general population the same? How are they different (RST.11–12.6, RST. 11–12.6, RST.9–10.6, RI.9–10.7)?

Cognitive Reflection

What types of careers involve assistive technology, prosthetics, and so on?

Keep, Retain, Generalize

There are many technologies in the *Star Wars* series. What kinds of technologies did you see in the series that exist now? Which ones do you think will be available in the next decade? Which do you think will never exist (RST.11–12.6, RST. 11–12.6, RST.9–10.6, RI.9–10.7)?

REFERENCES

Card, O. S. (2014). *Ender's game* (Vol. 1). Tor Teen.
Coogler, R. (Director). (2018). *Black panther* [Film]. Marvel Studios.
David, P. (2013). *After Earth: A novel*. Random House.
DeWitt, J., & Archer, L. (2015). Who aspires to a science career? A comparison of survey responses from primary and secondary school students. *International Journal of Science Education, 37*(13), 2170–2192.
Emran, A., Spektor-Levy, O., Tal, O. P., & Assaraf, O. B. Z. (2020). Understanding students' perceptions of the nature of science in the context of their gender and their parents' occupation. *Science & Education, 29*(1), 1–25.
Erduran, S. (2020). Nature of "STEM"? *Science & Education, 29*, 781–784. https://doi.org/10.1007/s11191-020-00150-6
Flegg, R. B., & Burke, C. (1995). The enigma of girls' concepts of the nature of science. *Australian Science Teachers Journal, 41*(3), 74.
Foster, A. D. (1978). *Splinter of the mind's eye*. Del Ray Books.
Henriksen, E. K., Dillon, J., & Ryder, J. (Eds.). (2015). *Understanding student participation and choice in science and technology education*. Springer.
Kamkwamba, W., & Mealer, B. (2010). *The boy who harnessed the wind*. HarperCollins.
McLeskey, J., Barringer, M. D., Billingsley, B., Brownell, M., Jackson, D., Kennedy, M., Lewis T., Maheady, L., Rodriguez, J., Scheeler, M. C., &, Winn, J. (2017). *High-leverage practices in special education*. Council for Exceptional Children.
Mesci, G., Schwartz, R. S., & Pleasants, B. A. S. (2020). Enabling factors of preservice science teachers' pedagogical content knowledge for nature of science and nature of scientific inquiry. *Science & Education, 29*(2), 263–297.
Meyer, S. (2005). *Twilight*. Little, Brown and Company.
Niccol, A. (Director/Screenwriter). (1997). *Gattaca* [Film]. Columbia Pictures.
Reilly, D., Neumann, D. L., & Andrews, G. (2019). Investigating gender differences in mathematics and science: results from the 2011 trends in mathematics and science survey. *Research in Science Education, 49*(1), 25– 50.
Riordan, R. (2005). *Percy Jackson and the Olympians: The lightning thief*. Miramax Books.
Sachar, L. (2008). *Holes*. Macmillan.

Schwartzbach-Kang, A., & Kang, E. (2018, February 22). *Using science to bring literature to life*. Edutopia. https://www.edutopia.org/article/using-science-bring-literature-life

Shelley, M. (2012). *Frankenstein*. Broadview Press.

Skloot, R. (2017). *The immortal life of Henrietta Lacks*. Broadway Paperbacks.

Trochim, W. M., & Donnelly, J. P. (2001). Research methods knowledge base.

Tsai, C. C., & Liu, S. Y. (2005). Developing a multidimensional instrument for assessing students' epistemological views toward science. *International Journal of Science Education, 27*(13), 1621–1638.

Vlahakis, G. N., Skordoulis, K., & Tampakis, K. (2014). Introduction: Science and literature special issue. *Science & Education, 23*(3), 521–526.

Wachowski, L., & Wachowski, L. (Directors) (1999). *The matrix* [Film]. Warner Bros.

Zeyer, A. (2018). Gender, complexity, and science for all: Systemizing and its impact on motivation to learn science for different science subjects. *Journal of Research in Science Teaching, 55*(2), 147–171.

CHAPTER 5

Mathematics and Literature

*Jugnu Agrawal, Diane Rodriguez, Gary Hoag, Ozalle Toms,
Ann Jolly, Elizabeth Reyes, and Ashley Voggt*

Professionals in special and general education as well as bilingual educators use some interesting methods, approaches, and strategies to teach mathematics to children. The hope is to keep all children motivated to achieve in mathematics and to even choose it as a vocation. There are some groups who might not choose mathematics as a field, such as females or culturally and linguistically diverse students (Raborn, 1995). Mkhize (2017) found that many students who choose not to pursue math and science curriculums are of average or above average intelligence; however, instruction still needs to be effective. What, then, constitutes effective instruction for these adolescents? Teachers will need to infuse methodologies that fit bilingual and general education students as well as students with special needs. How can teachers persuade them? Presenting mathematics in a novel manner sparks interest in many students and allows them to find out that they have a natural knack for mathematics. Integrating literature and mathematics is a viable approach that adds that touch of persuasion.

MATHEMATICS AND ADOLESCENTS

Adolescents desire independence; there is a strong need to engage with peers at that age, and learning differences can sometimes influence academic decisions. Many teachers believe that their students lack the cognitive ability to perform well in mathematics, that they do not possess adequate background in math fundamentals, or that they are not motivated. Sometimes students do not receive appropriate instruction and procedures are not scaffolded in a manner that enhances instruction. Mkhize (2017) reports that connecting mathematics instruction with motivational teaching results in the desired effect of encouraging the study of mathematics among adolescents.

Young females tend to choose socially oriented careers in fields such as health care or psychology, while boys often seek clinical research positions or medicine (Valenti et al., 2016). Mkhize (2017) proposes "action research in the classroom" that brings participation with other students.

McDonough and Ramirez (2018) found that teachers who are anxious about their own mathematic abilities can influence learners and can come to believe that math is far too difficult.

Beyond the gap in perceptions and self-determination, students with disabilities and culturally and linguistically diverse (CLD) students might struggle with reading math texts. Beal et al. (2010) note that performance on math assessments improves with performance in reading. The Lexile level of many texts may be too high for them, and they may lack the vocabulary they need to comprehend equations. Vocabulary study such as presented in this book is needed.

TEACHER APPLICATION TIPS

Teachers can use literature to teach various math concepts. Keollner et al. (2009) isolated adolescent literature that infuses math themes. They established three levels for literature; the higher the level of complexity, the greater the possibility for integrative efforts. Level 3 stories center on mathematical concepts; level 2 does not require understanding mathematical concepts but does require readers to understand how the story's central conflict was solved with math. Level 1 includes peripheral math so that readers can understand the story without much mathematical knowledge. Most of the books are at Keollner's level 1; therefore, teachers can locate literature that includes mathematical concepts at various levels depending on the needs and experience of the reader.

CAREERS

Mathematics is, among other things, a study of patterns. Each of the careers mentioned in this chapter involves a study of patterns, and understanding them can help our world in many ways. Furthermore, many careers that require math skills do not necessarily require a degree in math. Students are familiar with their favorite application on their phones; with a little expertise in computer science, math skills and, perhaps, a little specialized training in a couple of computer languages, students can develop apps themselves. Some careers include expertise with statistics (actuary, economist, statistician), while others are related to security (cryptographer). Security-related fields are increasingly important, and cryptographers who develop security system algorithms and cryptanalysts who decode and understand encrypted information are very much in demand. Some of the most common careers for those interested in mathematics include the following:

- **Actuaries** analyze the financial consequences of risk by using math and statistics.
- **Mathematicians** typically work to solve mathematical problems concerned with numbers, data, quantity, structure, space, models, and change.
- **Cryptographers** use algorithms and cyphers to encrypt sensitive data and are engaged in decrypting information.
- **Economists** research and think theoretically about financial problems and apply models to problems as they occur.
- **Statisticians** collect and analyze numerical data in large quantities, especially for assuming proportions as a whole from those in a representative sample.
- **Financial planners** assist people in managing risk and choosing appropriate investments.
- **Operations research analysts** provide insights for business decisions by helping people function more efficiently with data analysis, mathematical modeling, and a host of other skills.
- **Investment analysts** make certain that all known financial data is available for decisionmakers at critical moments.

LESSON PLANS

This chapter will have a number of literature books that infuse mathematic concepts. They are listed in Table 5.1 along with a summary and the skill that is taught.

Table 5.1. Summary of Literature Books/Films and Related Skills

Lesson 1 *To Kill a Mockingbird*	This story is about the author's observations of family near her hometown. The novel is known for depicting serious issues of racial inequality. The lesson focuses on statistics and probability.
Lesson 2 *The Hunger Games*	This novel is set in the future. Two "tributes" between the ages of 12 and 18 are chosen by lottery from the districts to compete in the televised "Hunger Games." The lesson focus is statistics and algebra.
Lesson 3 *Born a Crime*	This story is about a young man's relationship with his mother, who is determined to save her son from the cycle of poverty, violence, and abuse. The lesson focuses on writing a business plan.
Lesson 4 *Hidden Figures*	This book tells the true story of four African American female NASA mathematicians. The story captures their lives in the context of the Civil Rights Era, the Space Race, and the Cold War. Geometry and calculus are the skills practiced in this lesson.
Lesson 5 *The Giver*	Jonas is the Giver and he functions in a dystopian society. He holds everyone's memories. He fights to make change. The lesson focuses on the math in the pyramid.
Lesson 6 *The Chronicles of Narnia: The Lion, the Witch and the Wardrobe*	The story is a fantasy novel set in Narnia, a land of talking animals and mythical creatures that is ruled by the evil white witch. The lesson focuses on math and the weather.
Lesson 7 *Flatland*	The world of Flatland exists in two dimensions; all figures inhabit shapes, and the polygons have established themselves as the upper castes in society. The skill taught is geometry.

LESSON 5.1. STATISTICS AND PROBABILITY

To Kill a Mockingbird, by Harper Lee (1960)

Summary. *To Kill a Mockingbird* is a Pulitzer Prize–winning novel. The plot involves the author's observations of family and neighbors and an event that happened near her hometown. The novel is known for depicting serious issues of racial inequality. The narrator's father, an attorney, is the hero. Other primary themes of *To Kill a Mockingbird* are the destruction of innocence, courage, compassion, and gender roles in the South.

Application

Goal. Students will learn about probability and statistics while exploring race and inequality.

Objectives

- Given five assignments, students will research diaspora patterns with 90% accuracy.
- Given five words, students will be able to provide the meaning and be able to say and spell words with 100% accuracy.
- Students will be able to present information in graphs.

Standards. To view the Common Core Standards that correspond with this lesson, please visit the *STEAM Meets Story* page on www.tcpress.com and click on the Resources tab.

ACTIVITY 1

Teaching Strategy

Introduction. Students will analyze, synthesize, and evaluate multiple sources of statistical information presented in diverse formats in order to develop an understanding of different points of view regarding a period in history, and compare it to today in the United States (Math Content 6–8.SP.A.2, 6–8.SP.A.1) (Technology RI.6–10.7) (Science and Technology RST.6–10.3).

Materials

- Handouts
- Computers and tablets

Time. One to two days

Essential vocabulary. The words for sixth to eighth graders related to probability and statistics are listed in *Handout 5.1A.* The words are to be defined, the pronunciation is to be written, and a picture or example is to be provided (Science and Technology RST.6.-10.1, RST.6–10.5).

Trial/Try-Out

1. How many diverse groups were prevalent in the United States in the 60s compared to today? Were there any migration patterns of any significantly large ethnic groups (Math Content, 6–8.SP.A.1 and HSS.ID.A.1, 6–8.SP.A.2 and HSS.ID.A.2, 7.SP.B.3, 6–8.SP.B.4 and HSS.ID.A.4) (*Handout 5.1B*)?
   a. Locate two ethnic groups in the 60s and today that you would like to study. Is there a difference in population size presently compared to the past?
   b. In which states were the group most prevalent during the 60s? In what states are they presently predominately located?
   c. Are there different political structures within different states in the United States according to location? For example, are there differences in northern, southern, eastern, or western states?
   d. What can you infer about the two populations you are gathering data on and studying?
2. Present the data on the two diverse groups in charts, graphs, and other various forms. How many data sources did you use? Why is it important to use more than one data source (Math Content, 6–8.SP.A.1 and HSS.ID.A.1, 6–8.SP.A.2 and HSS.ID.A.2, 7.SP.B.3, 6–8.SP.B.4 and HSS. ID.A.4) (*Handout 5.1C*)?

Technology. Watch the series *Eyes on the Prize* (https://worldchannel.org/episode/eyes-on-the-prize-awakenings/?asset_slug=awakenings-1954-1956-t6dicm).

How do you think significant past events like the civil rights movement, or the Trail of Tears would be presented in the media today? Would various forms of technological communication make a difference? Do you think outcomes of these phenomena would be different? What about the Black Lives Matter movement? Is it the same or different than the civil rights movement? Why or why not (Literacy and Technology RL.6.7, RL.7.7, RL. 9–10.7,

RI.6.7, RI.7.7, RI.8.7, RI.9–10.7) (Science and Technology RST.6–8.9, RST. 9–10.9, RST, 6–8.8, RST 9–10.8)?

Additional Suggestions

1. On the diagram in *Handout 5.1D*, compare facts related to historical movements of two international, diverse groups. How are they like the U.S. diverse groups you studied? How are they different (Science and Technology RST.6–8.7, RST. 9–10.7)?
2. Have students read *To Kill a Mockingbird* (Literacy and Technology, RL.6–7.7, RL.9–10.7) (Science and Technology RST.6–8.5, RST.9–10.5).
   a. Discuss the book.
   b. Ask students to interview their parents and ask them about awareness of any similar events in the past.
   c. Why do you think the book has the name *To Kill a Mockingbird*? What does the mockingbird symbolize?

Assessment

Criteria for the previous assignments are listed in *Handout 5.1E*.

Cognitive Reflection

What other significant events involving diverse groups can you name?

Keep, Retain, and Generalize

What is the probability that the events in the story can happen today? Are similar events happening? What are they (Science and Technology RST.6–8.2, RST.9–10.2)?

ACTIVITY 2

Teaching Strategy

Introduction

1. In the story, Tom is tried and convicted even though Atticus proves that Tom could not have possibly committed the crime of which he is accused. Have a mock trial using the same characters in *To Kill a Mockingbird*, assuring a different outcome (Literacy and Technology RL. 6.7, RL. 7.7, RL.9–10, RI. 6.7, RI. 7.7, RI.9–10.7).

2. What is the probability of Tom winning at the time of the trial? What is the probability of Tom winning in present day (Statistics and Probability 7.SP.C.5, 8.SP.A.4, 6.SP.B.5, HSS.ID.B.5)?
3. Have students research and present a case where a person was proven innocent after being wrongly convicted of a crime. Techniques will include discussion, cooperative learning, and the writing process (Literacy and Technology RL. 6.7, RL. 7.7, RL.9–10, RI. 6.7, RI. 7.7, RI.9–10.7).

Materials

- Handouts
- Selected clips
- Copy of *To Kill a Mockingbird*
- Informational text

Time. Three to four class periods

Essential vocabulary. Look at the words in *Handout 5.1F*. Which words do you need to know? Use in context? Define? Spell (Literacy L.6.5, L.7.5, L.8.5, L.9–10.5)?

Trial/Try Out

For the trial, students are then to prepare briefs, opposing arguments, and evidence. Choose a prosecutor defendant, and judge. Is there a different outcome than in the book (*Handout 5.1G*) (Literacy and Technology RL. 6.7, RL. 7.7, RL.9–10, RI. 6.7, RI. 7.7, RI.9–10.7). (Literacy L.6.5, L.7.5, L.8.5, L.9–10.5)?

Technology. For the students' presentations on wrongful convictions, have them describe the arguments and evidence that led to their chosen research subject being exonerated (*Handout 5.1G*), and other contributions (Literacy and Technology R.L. 6–10.7) (Science and Technology RST. 6–8.6 and RST. 9–10.6).

1. What research methods will be used?
2. How will technology be included?

Additional Suggestions

1. Allow students to collaborate and work as a team. Make sure that students are able to participate using their strengths. What significance does the trial have as it relates to students with similar issues, such as those with disabilities or language differences? Have all students discuss, write, and

reflect (Literacy and Technology RL. 6.7, RL. 7.7, RL.9–10, RI. 6.7, RI. 7.7, RI.9–10.7) (Literacy L.6.5, L.7.5, L.8.5, L.9–10.5).

2. Read and discuss *To Kill a Mockingbird* (Science and Technology RST.6–8.2 and RST. 9–10.2).

3. Discussions should focus on incidents internationally or in other cultures/countries that have had some of the same challenges as non-majority populations in the United States.

Assessment

Measures of performance of the trial activities are listed in *Handout 5.1H.*

Cognitive Reflection

How does acquiring statistical information benefit your knowledge of various ethnic groups (Statistics and Probability 7.SP.C.5, 8.SP.A.4, 6.SP.B.5, HSS. ID.B.5)?

Keep, Retain, and Generalize

How does knowledge related to the cultural groups contribute to the future knowledge of statistics (Statistics and Probability 7.SP.C.5, 8.SP.A.4, 6.SP.B.5, HSS.ID.B.5)?

ACTIVITY 3

Teaching Strategy

Introduction. *To Kill a Mockingbird* takes place in the 1930s. Exploring matters from different perspectives and reading literature that involves characters from different cultures can help students become self-aware. Watch the film clips of Harriet Tubman and Harriet Beecher Stowe, analyze them in detail, and compare them to see how the early years of these two women contributed to them later becoming significant advocates for social justice (Science and Technology RST.6–10.7).

Materials

- Selected clips and resources for Harriet Tubman
  - » Film (https://www.youtube.com/watch?v =XmsNGrkbHm4)
  - » Books and resources (https://guides.loc.gov /harriet-tubman)

- Selected clips and resources for Harriet Beecher-Stowe
  - » Film (https://www.youtube.com/watch?v =ij Fy4RjYGbQ)
  - » Books and resources (https://www.history .com/topics/american-civil-war/harriet -beecher-stowe)
- Computers or tablets
- Copy of *To Kill a Mockingbird*
- Informational text

Time. One to two class periods

Trial/Try-Out

1. If you had the opportunity to meet with one of these characters today, (*Handout 5.1I*) (Science and Technology RST.6–10.10, RST. 6–10.8),
   a. who would you choose, and
   b. what questions would you ask?
2. Are there similar instances today that relate to the characters?
3. Write and act out a dialogue between two characters. Students can
   a. share dialogue, and
   b. create other characters.

Handout 5.1A. Activity 1: Essential Vocabulary

Define each word, write the pronunciation key, and draw or locate a picture.

Definition	Pronunciation	Picture
Probability		
Probability of independent events		
Fundamental counting principle		
Tree diagram		
Mean		
Median		
Range		
Bar Graph		
Line Graph		
Stem-and-leaf plot		
Circle Graph		
Histogram		
Scatterplot		

Handout 5.1B. Activity 1: Diverse Groups in the United States

Diverse Groups	States in Which They Are Most Prevalent	Migration Patterns From Other States and Countries	Statistical Changes in the Last 10 Years

Additional Suggestions

1. Examine the concept of respect in relation to varied cultures, gender, and disabilities.
2. Develop service learning projects related to diverse persons in the community.

Assessment

Handout 5.1J assesses creativity, language, setting, and effort.

Handout 5.1C. Activity 1: Statistical Media and Technology

In what varied ways are statistical data presented? Locate several statistical data sources related to African Americans that can be presented.

Source 1	Source 2
1.	1.
2.	2.
3.	3.
4.	4.
5.	5.

Handout 5.1D. Activity 1: Comparing Facts

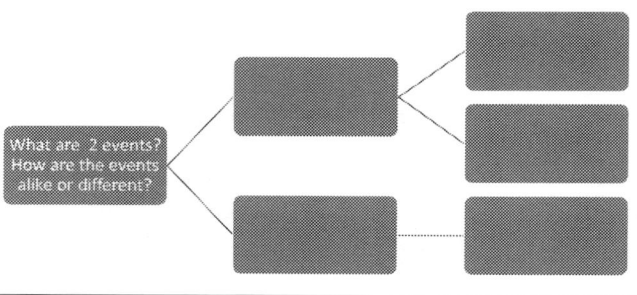

Handout 5.1E. Activity 1: Rubric for Statistics Media and Technology

	5 points	15 points	25 points
Varied sources of data listed	Two data sources	Three data sources	Five or more data sources
Presentation of data (charts, graphs)	No charts or graphs	Information represented by two charts or graphs	Three or more varied kinds charts of graphs are presented

Handout 5.1F. Activity 2: Essential Vocabulary, Trial Words

Define each word, write the pronunciation key, and draw or locate a picture.

Definition	Pronunciation	Picture
Circumstantial evidence		
Clerk of court		
Court reporter		
Cross-examine		
Defendant		
Evidence		
Hearsay		
Interview		
Oath		
Objection		
Oral argument		
Plaintiff		
Subpoena		
Trial		
Witness		

Handout 5.1G. Activity 2: Opening and Closing Arguments

What are the opening and closing arguments?
- ☐ Witnesses
- ☐ Facts

What evidence and research is needed?
- ☐ Positives
- ☐ Negatives

Cognitive Reflection

How would technology and the media affect the advocacy of Harriet Beecher-Stowe and Harriet Tubman today (Reading Informational Text RI.6–10.7)?

Keep, Retain, and Generalize

Narratives of people of various ethnicities and how they are affected by these events can be explored (Science and Technology RST.6–10.8).

Handout 5.1H. Activity 2: Rubric for Trial

	3 points	7 points	10 points
Presentation of argument	Appropriate argument points were not included.	Points were presented but a convincing argument was not present.	Appropriate points were presented and a strong argument was present.
Evidence	Evidence was weak and had few supports.	Evidence was presented but was not supportive to the incidents of the trial.	Strong evidence was presented.
Information and research	Very little research was presented and came from one source.	Research came from two to three sources.	Research came from four or more sources.
Statistical facts: science, medicine, art, sports, music	One category was presented.	Two categories were presented.	Three or more categories were presented.

Handout 5.1I. Activity 3: Role-Play

Role-play: Harriet Tubman and Harriet Beecher Stowe.

In an interview, what questions would you ask both characters?

Write a dialogue between the two characters.

Handout 5.1J. Activity 3: Rubric for Perspectives

	3 points	7 points	10 points
Creativity	Appropriate argument points were not included.	Points were presented but a convincing argument was not present.	Appropriate points were presented and a strong argument was present.
Language	Evidence was weak and had few supports.	Evidence was presented but was not supportive to the incidents of the trial.	Strong evidence was presented.
Setting	Very little research was presented and came from one source.	Research came from two to three sources.	Research came from four or more sources.
Effort	One category was presented.	Two categories were presented.	Three or more categories were presented.

LESSON 5.2. BASIC ALGEBRA AND STATISTICS

The Hunger Games, Suzanne Collins (2008)

Summary. This novel is set in the future. The nation of Panem consists of a capital that controls all resources, surrounded by 12 outlying districts, in the ruins of the area once known as North America. Teenage Katniss and her friend Gale, the main support for their families, hunt for food in the woods surrounding their impoverished district. They are approaching the annual reaping, when two "tributes" between the ages of 12 and 18 are chosen by lottery from the districts to compete in the Hunger Games, a televised survival contest where teenagers fight to the death.

Materials

- Group assignments
- Tablets with vocabulary words on Google Classroom
- Discussion questions
- Copies of *Hunger Games* text
- Appropriate handouts
- Selected video clips

Time. Two to three class periods

Application

Standards. To view the Common Core Standards that correspond with this lesson, please visit the *STEAM Meets Story* page on www.tcpress.com and click on the Resources tab.

Teaching Strategy

Introduction. Students work with two to three partners to discuss the following guiding questions for a class discussion:

- What does it mean to survive?
- How far would you go to survive?
- What images does the title of the novel bring to mind?

Essential vocabulary. Prior to the reading, especially for students with special needs and ELs, the following words will be pre-taught. These words were chosen because students may struggle with words that connect the *Hunger Games* and real-life world hunger. Students with disabilities or those who are at risk can make virtual flash cards using the WADE mnemonic to remember words and meanings. These can be posted on a vocabulary website (RL. 7.4, 8.2, 8.4, 8.9, 9–10.2, 9–10.4, 9–10.6) (*Handout 5.2A*).

Write the word and definition on a virtual flashcard.
Articulate the word using the dictionary pronunciation key.
Draw or find a digital a picture that reflects the meaning of the word.
Evaluate word usage by using it in a sentence.

Trial/Try Out

District 12 is one of Panem's poorest districts, but it's not the only district where children are starving. Katniss and Peeta, as the movie's protagonists, are very sensitive to the differences they see between the classes while on their Victory Tour, whereas citizens in the capital seem oblivious to the fact that people are starving and suffering in other districts. The sad truth is a large number of children are starving all over the world, and a considerable number of disadvantaged children are right here in the United States. According to the Children's Defense Fund (2021), more than 1.5 million children enrolled in public schools experienced homelessness during the 2017–2018 school year. The definition of food is quite the same, yet entirely different for the people of Panem. If you lived in the capital, food is simply a necessary luxury, and three meals a day come to you at a little cost. The majority of people in Panem labor just to get

Handout 5.2A. Essential Vocabulary

Word	Definition	Sentence
Teeming	Abundantly filled with especially living things	From this place, we are invisible but have a clear view of the valley, which is teeming with summer life, greens to gather, roots to dig, fish iridescent in the sunlight.
Drab	Lacking brightness or color; dull	Today her drab school outfit has been replaced by an expensive white dress, and her blonde hair is done up with a pink ribbon.
Smoldering	Showing scarcely suppressed anger	As we walk, I glance over at Gale's face, still smoldering underneath his stony expression.
Racketeer	Someone who commits crimes for profit	Most refuse dealing with the racketeers but carefully, carefully.
Sustenance	A source of materials to nourish the body	He lists the disasters, the droughts, the storms, the fires, the encroaching seas that swallowed up so much of the land, the brutal war for what little sustenance remained.
Torturous	Extremely painful or unpleasant	To make it humiliating as well as torturous, the Capitol requires us to treat the Hunger Games as a festivity, a sporting event pitting every district against the others.
Dissent	The act of protesting	So instead of acknowledging applause, I stand there unmoving while they take part in the boldest form of dissent they can manage.
Condone	Excuse, overlook, or make allowances for	We do not condone because it is wrong.
Threadbare	Thin and tattered with age	I had been in town, trying to trade some threadbare old baby clothes of Prim's in the public market, but there were no takers.
Keel over	Fall suddenly; collapse	I didn't pick it up for fear I would keel over and be unable to regain my footing.
Concoction	Any foodstuff made by combining different ingredient	But if I can hold down Greasy Sae's concoction of mice meat, pig entrails, and tree bark—a winter specialty—I'm determined to hang on to this.
Demeanor	The way a person behaves toward other people	She has dark brown skin and eyes, but other than that, she's very like Prim in size and demeanor.
Gnarled	Old and twisted and covered in lines	And some small gnarled place inside me hated her for her weakness, for her neglect, for the months she had put us through.
Tureen	Large deep serving dish with a cover	A tureen of fruit sits in ice to keep it chilled.
Tangible	Perceptible by the senses, especially the sense of touch	A hundred hands reach up to catch my kiss, as if it were a real and tangible thing.
Prestigious	Exerting influence by reason of high status	On the buildings that surround the Circle, every window is packed with the most prestigious citizens of the Capitol.

something to eat. If you lived in the poorer districts, food is never guaranteed and one meal a day can turn into none at all. Living in the district working, hunting, growing, or even stealing are all difficult, yet normal ways to obtain food. The advantage Katniss has over the people in the richer districts during the Hunger Games is that she knows how to hunt, specifically with a bow. The majority of people in our nation work to buy their own food, which is probably much better than the food served in the districts of Panem. There are a lot people in our world who hunt or grow their own food as well as suffer starvation.

1. The hunting game involves the force of the bow and arrow. How much force does it take to kill an animal with a bow and arrow? What distance do you have to stand from the animal? Do the calculations change in reference to the size of the animal (e.g., deer versus rabbit)? At what angle would you have to stand (Mathematic Calculations MC.7.EE.B.3, MC8F.B4, MC8F.B5, MC. 8.SPA.4, HSS.ID.B.5, HSS.ID.B.6., MC.HSS.IC.B.6, MC. 7.8P.C.8c.)?
2. With a team of two or three, create a poster session displaying statistics about childhood hunger in the United States according to the state or region

(Mathematic Calculations MC.7.EE.B.3, MC8F.B4, MC8F.B5, MC. 8.SPA.4, HSS.ID.B.5, HSS.ID.B.6., MC.HSS.IC.B.6, MC. 7.8P.C.8c.).

Assessment

Students' vocabulary will be assessed using Kahoot!. Kahoot! is a game-based learning platform used as educational technology in schools and other educational institutions. Its learning games, "Kahoots," are user-generated multiple-choice quizzes that can be accessed via a web browser or the Kahoot! app.

Cognitive Reflection

Find a college campus near you and participate in a poverty simulation (https://www.povertysimulation.net/) (MC.7.SP.C8.e).

Keep, Retain, and Generalization

In a team, decide on plans to have a food drive at your school. List the steps after you read the MEND article (https://mendpoverty.org/get-involved/volunteer-now/how-to-organize-your-own-food-drive).

LESSON 5.3. BUSINESS PLAN

Born a Crime, by Trevor Noah (2016)

Summary. *Born a Crime* is the story of a mischievous young boy who grows into a restless young man as he struggles to find himself in a world where he was never supposed to exist, given that his parents were subject to apartheid laws that prohibited interracial relations. It is also the story of that young man's relationship with his fearless, rebellious, and fervently religious mother—his teammate, a woman determined to save her son from the cycle of poverty, violence, and abuse that would ultimately threaten her own life.

Application

Goal. Students will learn how to create a business plan.

Objective. Given the correct tools, students will create a business plan with 90% accuracy using mathematics in accounting, inventory management, marketing, sales forecasting, and financial analysis.

Standards. To view the Common Core Standards that correspond with this lesson, please visit the *STEAM Meets Story* page on www.tcpress.com and click on the Resources tab.

Teaching Strategy

Introduction

1. After reading *Born a Crime*, students will be guided to design and create their own logical and mathematical advice. For example, Noah states, "My mother showed me what was possible." Noah's mother offers him advice and lessons throughout his life. Collect these lessons and pieces of advice from throughout the book and evaluate them. Which ones seem to mathematically benefit Noah most? Which ones do not? Collectively, what does his mother's advice help the reader understand about their relationship? How did this

advice impact his identity and sense of self (WHST.6–8.9, RST.9–10.7, L.6.5, L.7.5, L.8.5,L.9–10.5, RL.11–12.1)?

2. Noah has an epiphany when, about to sell a stolen digital camera, he looks at the pictures on it and has second thoughts. He reflects, "In society, we do horrible things to one another because we don't see the person it affects. We don't see their face. We don't see them as people. Which was the whole reason the hood was built in the first place, to keep the victims of apartheid out of sight and out of mind. Because if white people ever saw black people as human, they would see that slavery is unconscionable. We live in a world where we don't see the ramification of what we do to others, because we don't live with them" (pp. 221–222). What does this reflection suggest about the nature of guilt and how it influenced Noah's development and maturation? More broadly, what do his words suggest about the legacy of racism for South Africans? What is the impact economically? Evaluate why he anchors the book with the relationship with his mother and the impact of this decision (RL.11–12.5, RL.11–12.1, RL.11–12.2).

Materials

- Trevor Noah's *Born a Crime*
- Handouts
- Business plan outline
- Chromebooks with electronic spreadsheet software

Time. One to two class periods

Essential vocabulary. Students will collaborate with their partners to define vocabulary from the book: faction, futility, spate, encompass, sprawling, crony, expendable, robust, compendium, ramification, contrive, docket, recourse, manifestation. They will write these definitions in their notebooks to use in their business plan (L.6.5, L.7.5, L.8.5, L.9–10.5).

Trial/Try Outs

1. Students will be asked to explain "Make It Happen" (see *Handout 5.3A*). They will work with a partner and will be given visual examples of how they can design their super math concepts. Students will create a business plan and reflect on their business project (MC.7.EE.B.3, MC.8.F.B5, MCHSS.ID.B5).

2. Students will use an electronic spreadsheet to make calculations related to a business plan, which involves the entry of a few algebraic

expressions. To ensure that students have the requisite knowledge and skills to create a business plan in a spreadsheet, the lesson can be started with a review of the multiples of 5 (i.e., 5, 10, 15, 20, 25, 30, 35, 40, 45, 50, 55, 60) and the multiples of 6 (i.e., 6, 12, 18, 24, 30, 36, 42, 48, 54, 60, 66, and 72). Check that all students can create a spreadsheet that calculates the first twelve multiples of 5 and 6. This will ensure that students are able to enter an algebraic expression in a spreadsheet, which is a critical skill for success in this lesson (MC.7.EE.B.3, MC.8.F.B5, MCHSS.ID.B5).

In the first 12 rows of column A (the left-most column of the spreadsheet), students enter the numbers 1 through 12, as shown in Table 5.2.

Table 5.2. Structure of the Initial Spreadsheet

1		
2		
3		
4		
5		
6		
7		
8		
9		
10		
11		
12		

Table 5.3. Entry of Algebraic Expressions in Cells B1 and C1

1	=A1*5	=A1*6
2		
3		
4		
5		
6		
7		
8		
9		
10		
11		
12		

Next, in cell B1, students enter =A1* 5, and in cell C1 =A1 * 6, as shown in Table 5.3.

Immediately after the student enters =A1 * 5 into cell B1, the spreadsheet will multiply the value in cell A1, which is 1, by 5, and the result (5) will appear in cell B1. Similarly, since 1 * 6 is 6, the number 6 will appear in cell C1 immediately after the student enters =A1 * 6. This is shown in Table 5.4 and Table 5.5.

Table 5.4. Results After Entry of Algebraic Expressions

1	5	6
2		
3		
4		
5		
6		
7		
8		
9		
10		
11		
12		

Table 5.5. Results After Extending Algebraic Expressions

1	5	6
2	10	12
3	15	18
4	20	24
5	25	30
6	30	36
7	35	42
8	40	48
9	45	54
10	50	60
11	55	66
12	60	72

All that remains is for the student to extend the algebraic expression (formula) in cell B1 to cells B2 through B12. Then repeat that task in column C in order to extend the formula in cell C1 to cells C2 through C12. There are multiple ways to extend spreadsheet formulas in spreadsheets. For example, one could move the cursor to cell B1 and copy the formula by pressing Ctrl-C (on a computer running Microsoft Windows) or Command-C (on a Mac). Then move the cursor to cell B2 and paste the formula by pressing Ctrl-C (on a Windows PC) or Command-C (on a Mac). Subsequently, the formula can be pasted to the remaining cells by repeatedly moving the cursor and pasting the formula (for cells B3 through B12). The quicker way is to highlight all cells from B2 through B12 and paste (press Ctrl-C or Command-C) only once. Similarly, the formula in cell C1 is copied and pasted to cells C2 through C12. As an alternative to that copy-and-paste approach, a formula can be extended by dragging the rectangular handle in the lower right corner of a cell. Hence, with the cursor in cell B1, the handle in the lower right corner of cell B1 can be dragged to cell B12. Then move the cursor to cell C1 and drag the handle to cell C12. That's it! The values for the multiples of 5 and 6 are calculated and displayed automatically. Spreadsheet software automatically adjusts the formula by row, in this case to yield the multiples of 5 and 6. Given their knowledge of the multiples of 5 and 6, students can verify that they have correctly created the spreadsheet.

Students new to electronic spreadsheets should practice the entry of algebraic expressions as well as extending the expressions to subsequent rows and columns. One possible practice exercise, in light of the book's setting in South Africa, which uses the metric system, is to have students create a spreadsheet that converts kilometers to miles or kilometers per hour (kph) to miles per hour (mph). One need only multiply by 0.6 to convert kilometers to miles, or kph to mph. Hence, the student need only repeat the steps used to create column C, using 0.6 rather than 6. Additionally, rather than convert 1 through 12, students can be advised to put 10, 20, 30, 40, 50, 60, 70, 80, 90, 100, 110, 120 in column A in order to determine how many miles are in 10, 20, 30, . . . , 120 kilometers. Again, since students know the multiples of 6 and can divide by 10, many of them will be able to determine on their own whether they have created the spreadsheet correctly. Use of 0.6 as the constant to convert from kilometers to miles is perfectly fine for this application, given students' prior knowledge. For more precision in electronic calculations, one can use 0.6214 as the conversion constant.

Lastly, students will be asked to problem solve and create a business idea related to the music industry, using their spreadsheet skills to make calculations related to their hypothetical business. For instance, they might project revenue for 1 to 100,000 downloads of a song at costs of $1.00, $1.25, $1.50, $1.75, $2.00, $2.25, and $2.50 per download.

Students will then go on to use these calculations to design and implement a business plan (*Handout 5.3B*).

Handout 5.3A. Make It Happen

Write important match concepts.

Name creative ways to conduct a business plan.

Technology. Students will be allowed to use their Chromebooks and research music industry budgets. They can also use it to research the essential vocabulary.

Additional exercises. After the activity, students will be asked to write about why was it a huge risk to jump out of the taxi? Were there other choices Trevor Noah's mom could have made other than jumping out of the taxi with her children? What might have happened if she had made these other choices?

Writing and drawing: Make an illustrated timeline of the events of the story.

Outsider discussion: This is how Trevor Noah describes himself in grade 8: "Two things were true about me at that age. One, I was still the fastest kid in school. And two, I had no pride." How did Trevor use these things to his advantage? List two things that are true about you.

Writing: Imagine that Trevor was going to advertise his tuck-shop service at your school. What would be the best way for him to do this? Design an advertisement he could use to calculate cost.

Assessment

Using a teacher-created rubric, teachers will assess the business plan to see whether students have met fundamental objectives of the lesson. For example, have they established goals for each of the three areas? What is the multistep plan for achieving those goals?

Cognitive Reflection

Trevor Noah wrote of his experiences of discrimination and other hardships under apartheid. He highlights that we do not see the impact of our negative actions toward others. Reflect on the present condition of any group of persons of color and describe similarities and differences between challenges they face and the challenges Noah wrote about in his book.

Keep, Retain, and Generalize

Where else can you apply the knowledge gained through the creation of the business plan and spreadsheet that you learned in this lesson? Identify other careers where the use of a similar spreadsheet would be applicable

Handout 5.3B. Business Plan

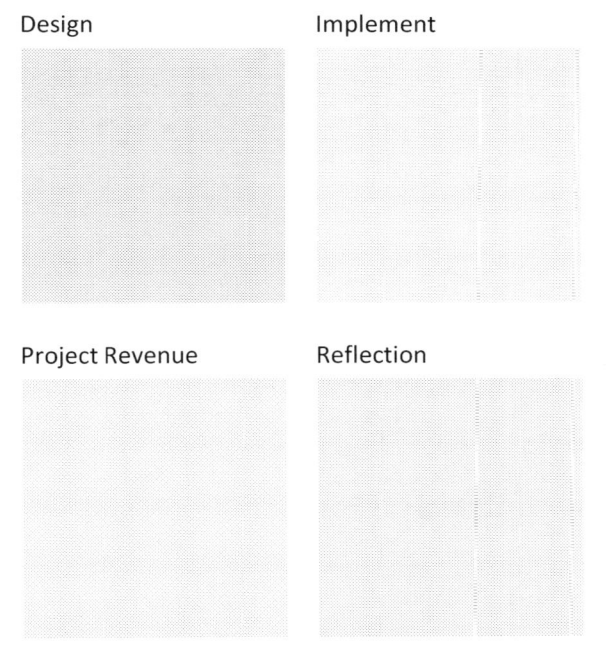

Design   Implement

Project Revenue   Reflection

LESSON 5.4. GEOMETRY AND CALCULUS

***Hidden Figures: The American Dream and Untold Story of the Black Women Mathematicians Who Helped Win the Space Race*, by Margot Lee Shetterly (2016)**

Summary. This book tells the true story of four African American female NASA mathematicians, Dorothy Vaughn, Mary Jackson, Katherine Johnson, and Christine Darden. These women used non-technological items, such as pencils, slide rules, and adding machines, to calculate figures for launching rockets into space. The story captures their lives in the context of the civil rights era, the Space Race, and the Cold War.

Application

Goal. Students will be able to determine how much fuel it takes to go to each planet in the solar system.

Objective. Students will compute to find how much fuel it might take to get from our planet to another.

Standards. To view the Common Core Standards that correspond with this lesson, please visit the *STEAM Meets Story* page on www.tcpress.com and click on the Resources tab.

Teaching Strategy

Introduction. It is important to note that the four figures discussed in the movie and the book were human computers (https://www.youtube.com/watch?v=qs42iZBVirA). They had to calculate without the use of a computer. Name the kind of calculations we can do now since we have computers that are not human.

Materials

- Script, PowerPoint with words and pictures, student vocabulary notebooks (e.g., composition notebook)

- Academic vocabulary for adolescents with learning disabilities. It is important to teach words explicitly, and systematically, because struggling readers and students with disabilities in learning may be less skilled than typical readers at gleaning word meanings.
- Collins Co-build Dictionary: https://www.collinsdictionary.com/
- Google Chrome or other accessibility tools
- Copies of *Hidden Figures: The American Dream and Untold Story of the Black Women Mathematicians Who Helped Win the Space Race*
- The Human Computer Project website: Project https://www.thehumancomputerproject.com/women

Time. Three to four 1-hour class periods

Essential Vocabulary

- Analysis
- Hypothesis
- Simulate
- Technical
- Transmit
- Trajectory
- Velocity
- Projection
- Variable
- Trajectory
- Velocity
- Fuel
- Angle

Trial/Try-Out

1. Explain words with pictures (select pictures representing a variety of culturally and linguistic diverse backgrounds) and their use in a student-friendly definition. Identify a synonym. Students may also reference a dictionary for additional support with definitions (CCSS.ELA-LITERACY.L.6.4.C; CCSS.ELA-LITERACY.RI.6).
   a. Sentence Writing: Provide opportunity for students to use vocabulary words in their writing
   b. Read text: *Hidden Figures* (CCSS.ELA-LITERACY.RI.6.1)
   c. Note taking: Students take notes (graphic organizer or stop and jot) as they read chapters/excerpts from text and discuss with peers and whole group.
2. The human computers calculated everything by hand (https://www.youtube.com/watch?v

=5wfrDhgUMGI&feature=emb_logo). Because the IBM computers were new, John Glenn did not want to depend on them and asked for "the smart lady" to calculate by hand. He was speaking of Katherine Johnson. Now that we have computers and calculators, in teams of four, see how far you can get in your research to do the following. We have a spaceship and we want to go to one of the other planets from Earth.

a. Make a model of the solar system on paper.
b. How far is each planet from Earth?
c. Can you find out how to calculate the trajectory from Earth to one of the planets?
d. Can you calculate how much fuel it would take to go from one planet to the other?
e. How did you go about finding your answer? Were you the human computer or did you use websites? If you used websites, which ones did you go to?

Assessment

1. Based on what you have learned about characteristics of astronauts and the human computers working at NASA (https://www.nasa.gov/audience/foreducators/topnav/materials/listbytype/OTM_Launch_It.html), describe the perfect human computer or astronaut. Draw a picture and a mock biography (be sure to include evidence from the text to support your work) (CCSS.ELA-LITERACY.RI.6.7; CCSS.ELA-LITERACY.RST.6–8.2; CCSS.ELA-LITERACY.RI.6.1; CCSS.ELA-LITERACY.RI.6.4; CCSS.ELA-

LITERACY.WHST.6–8.2.B). Students' work will be graded using *Handout 5.4A*.

2. Write questions for a mock job interview for an astronaut or human computer. What qualifications, skills, and knowledge must they have? Incorporate these characteristics into your questions. Write the interview in narrative form (CCSS.ELA-LITERACY.RI.6.7; CCSS.ELA-LITERACY.RST.6–8.2; CCSS.ELA-LITERACY.RI.6.1; CCSS.ELA-LITERACY.RI.6.4; CCSS.ELA-LITERACY.WHST.6–8.2.B).

Cognitive Reflection

Examine the learning that has occurred and how it connects to previous and future use of math skills we studied in the previous chapter.

1. How can you use social media or other means to advocate and encourage girls to go into STEM?
2. What classes or course path, in middle school and into high school, should students focus on to prepare themselves for a career in STEM?

Keep, Retain, and Generalize

1. Ask students, "Are there other professions or industries where you think similar patterns occur?"
2. Encourage students to develop a personalized learning network (PLN) and add reputable STEM organizations.

Handout 5.4A. Assessment Rubric (Low Inference)

Response includes	Below Expectations (1 pt)	Met Expectations (2 pts)	Exceeds Expectations (3 pts)
Evidence from texts	Less than one piece of evidence from text	Two or three pieces of evidence from text	Four or more pieces of evidence from text
Use of vocabulary words	One essential vocabulary word currently in context	Two or three essential vocabulary words used correctly in context	Four or more essential vocabulary words used correctly in context
Focus and structure	Limited organization; less than one paragraph	Clear organization; two or three paragraphs	Clear organization that builds logically; four or more paragraphs
Style (language and conventions	One or less words/phrases; more than 11 grammar and/or spelling errors	At least two or three transition words/phrases; between five and 10 grammar and/or spelling errors	Four or more transition words/ phrases; four or less grammar and/or spelling errors

LESSON 5.5. MYSTICISM OF THE PYRAMIDS

The Giver by Lois Lowry (1993)

Summary. *The Giver* is a story of a boy named Jonas who lives in a society that is being controlled by the rules and tradition of the elders. Jonas is selected as the Receiver of Memory, a position that distinguishes him from others and gives him authority. But with this authority comes tremendous responsibility.

Application

Goal. Students will learn vocabulary words and will learn about ancient Egyptian mathematics.

Objectives

- Students will be able say and define at least five vocabulary words.
- Students will be able to research and perform at least two Egyptian measures.

Standards. To view the Common Core Standards that correspond with this lesson, please visit the *STEAM Meets Story* page on www.tcpress.com and click on the Resources tab.

Teaching Strategy

Introduction. *The Giver* is set in a future dystopian society. The book starts with the small town as a utopia, but it dissolves. One person is given knowledge and memories, and it is kept from the others. The way the community handles sick babies is to exterminate them, but only the Giver knows that. Jonas, the Giver, wants to find an escape with the infant who should have been with his family. What would you consider a utopia?

Materials

- *The Giver* (book/audiobook and 2014 film)
- Prezi to present partner summaries by chapter

Time. Two to three class periods

Essential vocabulary (RST. 6–8.4, RST. 9–10.4, L.6.8, L.7.5, L.8.5, L.9–10.5).

- Memories
- Hieroglyphics

- Ancient Egypt
- Pythagorean theorem
- Pyramid
- Utopia
- Dystopia
- Protest

Task: Students should (a) determine word meanings from context, (b) verify meaning in an online dictionary, and (c) create online flash cards on Quizlet.

Trial/Try-Out

In the book and movie, the students go inside a pyramid (https://www.youtube.com/watch?v=DIETtH8IJoQ). Pyramids are surrounded by mathematical concepts and mysticism. The ancient Egyptians were mathematicians. They measured time, surveyed land, determined the Nile flooding, and developed a monetary trade scheme. They created the first calendar and built the pyramids. The Egyptians also employed original multiplying, dividing, and fraction techniques. Everything was multiplied by 2, so if they wanted to find some quantity, say x multiplied by 5, they would turn that into $x * 2 + x * 2 + x$. Regarding division, 13/2 would be done by $4 * 2 + 4 = 12$, $13 - 12 = 1$, and so the answer was 6.5. When the ancient Egyptians wrote whole numbers, like 32, they would have to write $10 + 10 + 10 + 1 + 1$ because those were the symbols they had created in hieroglyphics. The Egyptians also used a rope knotted into 12 sections that stretched out to form a 3-4-5 right triangle, thus utilizing the Pythagorean theorem prior to the ancient Greeks (RST. 6–8.9, RST.9–10.9, RST-11–12.9).

1. What is the Pythagorean theorem? Why is it used?
2. With a team, try the varied ways the Egyptians multiplied, divided, and used fractions. Did it work?
3. Can you find research on the knotted rope and its uses?
4. Can you find information about the ways of the ancient Egyptians?
5. What was mystical about the pyramid in *The Giver*?
6. Students will present their findings using Prezi.

Assessment

Rubric for evaluating student presentations is 100 points.

- Is the information researched correct (25 points)?

- Are the vocabulary words used appropriately (25 points)?
- Did the team work well together (25 points)?
- Is the presentation well thought through (25 points)?

Cognitive Reflection

What are the pros and cons of rules and laws for society? Are there any being broken? Are there groups protesting? Why are they protesting? Do you believe they are right or wrong? Did Jonas protest? What did he protest? Discuss (RST. 6–8.9, RST.9–10.9, RST-11–12.9).

Keep, Retain, and Generalize

1. Is what you consider a utopia anything like the world you live in now? How is it different (RST. 6–8.9, RST.9–10.9, RST-11–12.9)?
2. How different is current Egypt? What are the differences? How did it get that way? Do you believe the people there are mathematicians (RST. 6–8.9, RST.9–10.9, RST-11–12.9)?

LESSON 5.6. WEATHER MATHEMATICS

The Chronicles of Narnia: The Lion, the Witch and the Wardrobe
by C. S. Lewis (1950)

Summary. *The Lion, the Witch and the Wardrobe* is a fantasy novel set in Narnia, a land of talking animals and mythical creatures that is ruled by the evil white witch. Four siblings find themselves adventuring to Narnia to save their own lives. The lion Aslan gives his life to save one of the children and then later rises from the dead to vanquish the white witch and crowns the children kings and queens of Narnia. The novel teaches about the themes of betrayal and forgiveness because one of the characters betrays his own family in exchange for promise of power.

Application

Goal. Students will be able to determine that meteorology is a math field.

Objective. Given mathematical questions involving weather, students will be able to research the information.

Standards. To view the Common Core Standards that correspond with this lesson, please visit the *STEAM Meets Story* page on www.tcpress.com and click on the Resources tab.

Teaching Strategy

Introduction. Peter, Susan, Edmund, and Lucy Pevensie are sent to live in Professor Kirke's home, along with his **housekeeper**, Mrs. Macready, in the English countryside during World War II. As the children **investigate** the house, Lucy discovers an old wardrobe in a **spare** room. The wardrobe is actually a passage to Narnia, a world filled with magic. Lucy goes through the wardrobe and meets a goat-legged man (a faun) named Mr. Tumnus. She learns that Narnia is ruled by the evil white witch who keeps Narnia under an **eternal** winter. When Lucy returns to Professor Kirke's house, she **realizes** that though she spent hours in Narnia, no time has passed in her world. Her siblings do not believe her story about Narnia because the wardrobe's **portal** doesn't work when they try to go through it. A few days later, when the children are playing hide-and-seek, Lucy hides in the wardrobe, and Edmund follows her into Narnia. Edmund meets the white witch, who introduces herself as the queen of Narnia. The witch feeds Edmund an enchanted form of **Turkish Delight** and asks him to bring his brother and sisters. Where are the four Pevensie children sent to live during World War II? Why does the wardrobe send them to another dimension? Why do you think the white witch wants Edmond to bring his siblings into Narnia? Why do you think the weather is different in Narnia?

Materials

- Copies of *The Lion, the Witch and the Wardrobe* (book 1950/film 2013)
- Prezi to present character analysis and general summary
- Accessible technology to get online (iPad, Chromebook, computer, etc.)

Time. One to two class periods

Essential vocabulary: Meteorologist, prediction, forecast (L.6.5, L.7.5, L.8.5, L.9–10.5)

Trial/Try-Out

1. Immediately when the children went through the wardrobe to Narnia in another dimension, they had to put on coats. It was a very cold place. You find out later on that the long winter came to a dramatic end magically. The weather can change dramatically in our dimension also. What really strange weather conditions have you seen? Tornadoes and hurricanes require skilled people to analyze, report, and predict the weather. These are the kinds of jobs you can have related to predicting the weather:
   - An operational forecaster analyzes weather conditions and issues forecasts or alerts the public of severe weather for their area.
   - A research meteorologist studies more specific areas of weather like severe weather or climate change. They can also develop tools like radar or weather models to help other meteorologists in their jobs.
   - A meteorologist in the military makes weather observations and forecasts for missions around the world.
   - Airlines use meteorologists to help pilots know what the weather will be like during take-off, in flight, and landing.
   - Electric companies ask meteorologists about extreme cold and heat.
   - Did you know that observing weather is a math skill? How is it a math skill?
2. Observing the weather requires math skills. With a team, examine the math skills needed to study adverse weather conditions such as a hurricane. What is the math behind the water temperature, moisture content of the air, barometric pressure, and movements of wind and currents? A hurricane is very destructive; how do you calculate the power of the hurricane? How is wind speed determined? How do you determine the category of a hurricane? What is an infamous hurricane? Compare it to a less infamous one. Compare the size, the power, and damage of both hurricanes (e.g., damaged homes, human lives, number of states, flooding, number of people displaced). Can you describe a hurricane's effect on you or someone you know (MC.HSS.IC.B.4, MC.HSS.IC.B.5, MC.HSS.IC.B.6)?

Assessment

Rubric for evaluating student presentations is 100 points:

- Is the information researched correct (25 points)?
- Are the vocabulary words used appropriately (25 points)?
- Did the team work well together (25 points)?
- Is the presentation well thought through (25 points)?

Cognitive Reflection

Discuss if Edmond's character is seduced by the idea of power in Narnia and willing to betray his family. Reflect on a current movie/TV show where a character has done something similar and tell how the two are similar (RST.6.8.9, RST.9–10.9).

Keep, Retain, and Generalize

How does betrayal typically work out for the person who betrays? What are some consequences of betraying someone? Justify the choices you made and the thought process that went into decisionmaking.

LESSON 5.7. GEOMETRY

Flatland by Edwin Abbott (1884)

Summary. The book anticipates many of the dystopian tales and events of the 20th century. The world of Flatland exists in two dimensions; all figures inhabit shapes, and the polygons have established themselves as the upper castes in society.

Application

Goal. Students will be able to demonstrate understanding of geometrical shapes including triangle, pentagon, hexagon, polygon, and circle, as well as patterns.

Objectives

- Students will answer questions and perform calculations with 90% accuracy.
- Students will demonstrate the value of patterns and create an algebraic equation to solve a real-life problem.

Standards. To view the Common Core Standards that correspond with this lesson, please visit the *STEAM Meets Story* page on www.tcpress.com and click on the Resources tab.

ACTIVITY 1

Teaching Strategy

Introduction. Students will watch the trailer of the film *Flatland* (https://www.youtube.com/watch?v = C8oiwnNlyE4). Students will engage in whole class discussion using the prompt of defining congruence. What role did it play in providing an understanding of the universe for the inhabitants of Flatland? What conflicts appear? How are those conflicts related to geometry? What does the external conflict have to do with dimension (HGS.CO.A.1–5)?

Essential vocabulary. In addition to geometric terms, students will master use of the following terms: congruence, dimension, geometry, configuration, chromatic, dimension, infer, chromatist, configuration, circle, circumference, perspective. Students will create morpheme cards for morphemes in each word (HGS.CO.A.1–5).

Materials. Students use the morpheme card template to complete the vocabulary exercise. They will also need paper, scissors, and glue to build shapes from printable online templates (https://www.polyhedra.net/en/).

Trial/Try-Out

1. Students pair up to solve the problems and write down their answers to the following prompt: When the protagonist (square A) says to his son that he must infer what the light tells him about various shapes, what does he mean? Students use forms and instructions at https://www.polyhedra .net/en/ to create geometric shapes. They shine a light (from their phone or a small flashlight) to observe how light and shadow contrast. What is the difference between a square and a triangle? A triangle and a hexagon? What are your observations (HGS.CO.A.1–5)?

2. Review the following resources related to Flatland: https://ned.ipac.caltech.edu/level5 /Abbott/paper.pdf; https://www.youtube.com /watch?v=MGv8MMi8QO0; https://www.youtube .com/watch?v=eyuNrm4VK2w. Students will watch the TED Talk on *Flatland* as well as the film; they will use the book as a reference. They will engage in whole class discussion using the following prompt: Do the themes relate to events of the 20th century and beyond (HGS.CO.A.1–5)?

3. Students will break into groups using the Stand up, Hands up, Pair up cooperative working techniques and complete the attached exercise on the Flatland caste system (*Handout 5.7A*). As students listen to other groups, they will write conclusions on Padlet, noting their reaction to the presentation and responding to the following questions:
   - Why is the possibility of a third dimension shocking to inhabitants of Flatland?
   - Why does the protagonist finally disagree with the leaders of Flatland?
   - How does the book and film relate to the 20th century and to contemporary time (HGS. CO.A.1–5)?

Assessment

Formative. Do students recognize geometric terms?

Summative. Are students able to relate geometric terms to other literary terms or story events?

Cognitive Reflection

Whole class discussion should take place after the presentations of the questions listed.

Exit ticket. Students respond to the prompt in writing on Padlet: What internal conflict does the protagonist confront? Is the protagonist reliable or unreliable? Students can use hard-copy Exit Ticket as an alternative (HGS.CO.A.1–5).

Know, Retain, and Generalize

Did this lesson given you new ideas about geometry? What are they? Has this lesson on Flatland given you new perspectives on questions of fact versus fiction? What are those perspectives (HGS.CO.A.1–5)?

Handout 5.7A. Activity 2: The Caste System of *Flatland*

What's a caste system? In the film and book *Flatland,* a system of castes is referenced as a fact of life. Elites have fostered castes in many human societies. Such castes generally segregate people by race, color of skin, and socioeconomic status. In Flatland, castes are based on shape. Women, who are all just straight lines, are at the low end, while polygons and circles are at the top. Some polygons defy easy classification, and the government often sends them for reconfiguration, which means they are changed to fit the next lowest official shape.

Following are a series of shapes. Someone has put them out of their "natural" order. Your job is to write each shape's name in column 1 and place it in the correct order from top to bottom according to the Flatland's caste system in column 2. One polygon does not fit the order; match it with reconfiguration.

Name of the Shape	Shape	Column 1: Name of Shape	Column 2: Order in the Caste System (High to Low)
Hexagon			
Triangle			
Line			
Pentagon			
Polygon			
Reconfigure			
Square			
Rectangle			
Circle			

ACTIVITY 2

Teaching Strategy

Introduction

1. Students will engage in whole class discussion using the following prompt: Do the themes in the text relate to events of the 20th century (HGS. CO.A.1-5)?
2. Students will divide into groups of three and create story maps that identify the role of each literary term in the story of Flatland (6.NS.B.3, 6.EE.A.1, 6.EE.A.2.C).

Essential vocabulary. Students will master use of the following vocabulary terms: chromatist, configuration, circle, circumference, infinite, apostle, perspective, sphere, and solid.

Each student will be assigned to research a vocabulary word to collect the following information.

- Origin of the word (www.etymonline.com)
- Part of speech
- Definition of the morphemes
- An illustration
- Use the vocabulary word in a sentence

Students will present the assigned word to the class.

Trial/Try-Out

Students will complete their story maps and present them to the class. As students listen to other groups, they will write conclusions on Padlet, noting their reaction to the presentations and responding to the following questions:

1. How does the protagonist respond to the discovery of a third dimension?
2. What does the protagonist say to the monarch of Pointland (the land of the first dimension)?
3. What is the fate of the protagonist? What does his fate say about the future of tolerance (HGS.CO.A.1–5)?

Assessment

Formative. Do students understand basic geometric concepts?

Summative. Are they able to discuss the relevance of various geometric figures and their importance to the story?

Cognitive Reflection

Exit ticket. What does the protagonist's fate say about the future of tolerance in human societies (remember, the book was originally authored in 1884)?

Know, Retain, and Generalize

Has this unit given you new perspectives on questions of fact versus fiction? What are those perspectives (HGS.CO.A.1–5)?

REFERENCES

American Institutes for Research. (n.d.). *National Center on Family Homelessness*. https://www.air.org/center/national-center-family-homelessness

Beal, C. R., Adams, N. M., & Cohen, P. R. (2010). Reading proficiency and mathematics problem solving by high school English language learners. *Urban Education, 45*(1), 58–74. https://doi.org/10.1177/0042085909352143

Collins, S. (2008). *The hunger games*. Scholastic Press.

Keollner, K., Wallace, F. H., & Swackhamer, L. (2009). Integrating literature to support mathematics learning in middle school. *Middle School Journal, 41*(2), 30–39.

Lee, H. (1960). *To kill a mockingbird*. J. B. Lippincott & Co.

Lewis, C. S. (1950). *The chronicles of Narnia: The lion, the witch and the wardrobe*. Geoffrey Bles.

Lowry, L. (1993). *The Giver*. Houghton Mifflin Harcourt.

Mkhize, D. R. (2017). Forming positive identities to enhance mathematics learning among adolescents. *Universal Journal of Educational Research, 5*(2), 175–180. https://doi.org/10.13189/ujer.2017.050201

McDonough, I. M., & Ramirez, G. (2018). Individual differences in math anxiety and math self-concept promote forgetting in a directed forgetting paradigm. *Learning and Individual Differences, 64*, 33–42. https://doi.org/10.1016/j.lindif.2018.04.007

Noah, T. (2016). *Born a crime*. Spiegel & Grau.

Raborn, D. T. (1995). Mathematics for students with learning disabilities from language-minority backgrounds: Recommendations for teaching. *New York State Association for Bilingual Education Journal, 10*, 25–33.

Shetterly, M. L. (2016). *Hidden figures: The American dream and untold story of the Black women mathematicians who helped win the space race*. HarperCollins.

Valenti, S. S., Masnick, A. M., Cox, B. D., & Osman, C. J. (2016). Adolescents' and emerging adults' implicit attitudes about STEM careers: Science is not creative. *Science Education International, 27*(1), 40–58.

CHAPTER 6

Engineering and Literature

*Diane Rodriguez, Gloria Campbell-Whatley, Kevin Otero, Hassan Payano,
Maria Payano, Eileen Interiano, and Sharon Hunter*

The need for more skilled workers in engineering has been observed for the last several decades, and the shortage of this talent has become obvious (National Science and Technology Council [NSTC], 2018). New generations of engineering workers can help solve problems of climate pollution by developing environmentally friendly systems through real-world applications of engineering-based learning (Wu et al., 2019). To meet the increasing demands of our country and world, the NSTC (2018) and the U.S. Department of Education (2018) have called for an improvement of teaching engineering to help better prepare our next generation of pioneers.

ENGINEERING FOR ADOLESCENTS

Engineers who design technologies use an understanding of science and math and provide an adhesive context of science and math together. Most adolescents don't know what an engineer does, which implicates a nationwide fundamental lack of understanding of the field. Familiarizing and producing excitement about engineering in the adolescent grades with innovative teaching techniques is a challenge. There is a need to allocate instructional time focused on the sciences (Mantzicopoulos et al., 2009; Romance & Vitale, 1992). The challenge can be met by linking engineering with literature instruction and using math and science to solve problems in a situational context. Adolescents have to develop an engineering identity because of the known relationship between a student's academic identity and their goals (Was et al., 2009). Developing mathematics and science identities contributes to motivation and persistence in engineering fields and develops a learning context for scientific academic identities (e.g., Berry et al., 2011). Therefore, exposing students to engineering activities can support the development of positive engineering identities.

Radunzel and colleagues (2016) challenged educators and policymakers to ensure engineering students are prepared for rigorous careers. This can be done by ensuring that elementary and high schools are teaching the highest level of STEM. High school courses should include physics, chemistry, and calculus.

In most science fields, including engineering, there is a significant achievement gap among female and male participants. Shi (2018) indicated that high school academic preparation contributed significantly to the gap in SAT scores and grade point averages (GPAs). Wei and colleagues (2017), found that adolescents with mild disabilities who took more advanced-level STEM classes were more likely to declare a STEM focus or an engineering major in college, especially courses in calculus and science that are associated with the engineering pipeline.

Students from culturally and linguistically diverse (CLD) backgrounds have even more challenges. When diverse students are screened, engineering students from underrepresented communities may be overlooked for their talents because traditional instruments such as IQ and achievement tests may not reveal the talent and motivation CLD adolescents may possess (Wu et al., 2019). An alternative approach to the identification of talented STEM students is performance-based assessments, including open-ended assignments that encourage higher order thinking and problem solving (Wu et al., 2019). The vital nature of mentoring cannot be overstated in engineering students' development. The learning environment of an engineering program is critical to the success of CLD adolescents (LaForest et al., 2020; Shi, 2018; Spillane et al., 2016; Wu et al., 2019). Mentors help to improve performance and increase self-efficacy (Morales, 2010; Wu et al., 2019) of CLD populations. Precollege programs cultivate engineering development (Wu et al., 2019) and nurture young engineering talent in CLD populations (Spillane et al., 2016; Wu et al., 2019).

CAREERS

Jennings (2015) conducted a study that explored young adolescents' interest in engineering as a future career. Students viewed a brief engineering video and wrote why they felt the same or different about engineering following the video. Overall, adolescents were more positive about engineering as a field, and many myths about engineering were dispelled. Qualitative analyses revealed that females more than males noted that engineers helped people. The video included both men and women as engineers, and girls had more positive responses about the field of engineering as a future career.

It is important to expose adolescents to different careers to provide opportunities to explore future paths. The field of engineering includes careers in chemical engineering, civil engineering, electrical engineering, and mechanical engineering. These areas can be introduced to adolescents to let them know what the field of science can offer them. Some careers in engineering include but are not limited to the following:

- *Chemical engineers* influence various areas of technology by conceptualizing and designing processes for producing, transforming, and transporting materials.
- *Civil engineers* design and supervise large construction projects, including roads, buildings, airports, tunnels, dams, bridges, and systems for water supply and sewage treatment.
- *Electrical engineers* design and develop new electrical equipment, solve problems, and test equipment.
- *Mechanical engineers* design, develop, build, and test mechanical devices, including tools, engines and machines.
- *Aerospace engineers* design aircraft, spacecraft, satellites, missiles, and systems for defense.
- *Biochemical engineers* analyze and design solutions to problems in biology and medicine, with the goal of improving the quality and effectiveness of patient care.
- *Computer hardware engineers* research, design develop, and test computer equipment such as chips, circuit boards, or routers.
- *Computer systems engineers* combine their knowledge of computer science, engineering, and mathematical analysis to develop, test, and evaluate software, circuits, personal computers, and more.
- *Environmental engineers* use the principles of engineering, soil science, biology, and chemistry to develop solutions to environmental problems.
- *Flight engineers* are responsible for ensuring that all components of the plane are in proper working order.
- *Industrial engineers* develop solutions to eliminate wastefulness in production processes. They devise efficient ways to use workers, machines, materials, information, and energy to make a product or provide a service.
- *Marine engineers* design, build, test, and repair ships, boats, underwater craft, offshore platforms, and drilling equipment.
- *Mechatronics engineers* are uniquely equipped to work as mechanical engineers with electronics, instrumentation, and real-time software engineering skills.
- *Mining and geological engineer* mine for the safe and efficient removal of minerals (such as coal and metals) for manufacturing and utilities.
- *Nanotechnology engineers* test for pollutants, create powders to enrich our foods and medicines, and study the smallest fragments of DNA.
- *Nuclear engineers* develop the processes, instruments, and systems used to get benefits from nuclear energy and radiation.
- *Robotics engineers* are behind-the-scenes designers responsible for creating robots and robotic systems that are able to perform duties humans are either unable or prefer not to complete.
- *Software engineers* are engaged in computer software development and apply engineering principles to software creation.
- *Hydro/hydraulic/water engineers* provide clean water, dispose of waste water and sewage, and prevent flood damage.

TIPS FOR TEACHERS

When students hear the word *engineer*, they may not view it as a broad term covering many areas, but it covers a wide range of skills. Engineers not only design, build, and maintain machines but also help create everything around us. Engineers use a design process, which is a step-by-step scaffolding process to solve complex problems. Teachers should help students understand that an engineer is behind many topics they may want to explore. They should ensure that students work as teams, just as engineers do, and start with the easier activities first and later move to more advanced activities that require science and math. This book is integrated with many interesting and novel topics. Make sure that students have fun while they learn and teach them in such a way that it peaks their interest. Students at risk for learning disabilities and students with disabilities might amaze you with their ability to perform cognitively because the lessons are highly interesting. Make sure to encourage their involvement and engagement. The lessons are well integrated, with numerous practice opportunities threaded throughout. Practice and exploration can be woven into these lessons.

SUMMARY OF LESSONS

It is important that teachers integrate high-interest texts such as *Maze Runner*, *Harry Potter*, and *Ender's Games* to teach engineering concepts. For example, these stories contain riddles, adventures, and problem-solving that can encourage ingenuity in engineering. They are similar to video games of today in the adventures that youth experience (Schwartzbach-Kang & Kang, 2018). These stories have been incorporated into the following lessons to promote various engineering concepts (see Table 6.1).

Table 6.1. Summary of Literature Books/Films and Related Skills

Book	Summary and Engineering Skills
Lesson 1 *Maze Runner*	In the *Maze Runner*, Thomas wakes up trapped in a glade that is surrounded by stone walls and doors that lead to a maze. He travels the maze to find a means to escape. The activities in the lesson focus on robotics.
Lesson 2 *Harry Potter* series	*Harry Potter* is a series of novels about magic. The incidents occur at the Hogwarts School of Witchcraft and Wizardry. Construction and building skills are highlighted in this lesson.
Lesson 3 *Men in Black*	Men in Black is a top-secret agency committed to pursuing and monitoring the activities and actions of aliens. The lesson focuses on drones.
Lesson 4 *Snowpiercer*	Survivors of Earth's second Ice Age live out their days on a luxury train that ploughs through snow and ice. This lesson centers on various train engines.
Lesson 5 *Star Trek—The Motion Picture: A Novel*	The Starship Enterprise has a mission to find a power that threatens all mankind on the Earth. The Voyager 6 satellite has come back to meet the creator. Satellite engineering is the crux of this lesson.
Lesson 6 *The Terminator*	A cyborg assassin known as the Terminator travels from 2029 to 1984 to kill Sarah Connor, Kyle Reese is sent by Sarah's son to protect her. This lesson centers on cyborgs and their importance in present day.
Lesson 7 *Jumanji*	*Jumanji* is about children who discover a strange game. In the 2017 film based on the book, they are sucked into the setting and become part of the game's software. This lesson focuses on software engineering.
Lesson 8 *Minority Report*	*Minority Report* is an action-detective thriller set in 2054. Police use specialized mutated humans, called precogs, to predict murders before they happen. The lesson focuses on aerospace engineering.
Lesson 9 *The Martian*	In 2035, the National Aeronautics and Space Administration (NASA) sent a man to Mars for 1 month. A large dust and windstorm appeared, and he was trapped. He must rely on his own resourcefulness to survive. The lesson introduces space travel.
Lesson 10 *Back to the Future*	A school-aged student and a scientist attempt to get fuel for time travel in a car. This lesson focuses on biofuels.

LESSON 6.1. ENGINEERING ROBOTICS

The Maze Runner by James Dashner (2013)

Summary. *The Maze Runner* is a series of action-filled books in which Thomas and his friends experience unbelievable trials. In the first book, Thomas and friends wake up stuck in a glade that is surrounded by stone walls and doors that close each night to protect them. Each day, boys are sent through the maze to find a means of escape. There are creatures that attack them in the maze. When the boys figure out the secret to escape the maze, they are sent to *The Scorch Trials*, which is the second book of the series. They have to cross the hottest climate on Earth in 2 weeks or they will die of a viral infection. Thomas and his friends must beat the odds to survive. *The Death Cure* is the last book in the series. In this book, Thomas and his friends fight WICKED, the organization that designed the entire test.

Application

Goal. Students will understand the rudiments of robotics.

Objectives

- Students will write directions for the maze as a practice for programming a simple robot.
- Students will be able to identify robots on a construction worksite.
- Students will be able to identify uses of cyborg (tissue and machinery combined) type technology.

Standards. To view the Common Core Standards that correspond with this lesson, please visit the *STEAM Meets Story* page on www.tcpress.com and click on the Resources tab.

Teaching Strategy

Introduction. Students are introduced to the logic necessary to solve a maze puzzle. First, they observe a blindfolded student volunteer to be guided through a classroom maze using the verbal directions of another student. In this demonstration, the blindfolded student represents a robot and the guiding student represents programming commands. Programming logic is applied using a robotic kit to navigate through a maze (Science and Technical Subjects, RST.6–8.3).

Materials

- Handouts
- LEGO robots, OWI robotic kit, Smart Machines robotic kit, or another
- Copy of the book *The Maze Runner*
- Classroom informational text
- Large white bed sheet or a large sheet of paper to draw the maze
- Measuring instruments

Time. One to two class periods

Essential vocabulary. Write the meaning of the following words. Collaborate with a classmate using flash cards or virtual cards to pronounce and define the words and say the meaning orally (Science and Technical Subjects, RST. 6–8.4).

- Design
- Logic
- Maze
- Programming
- Robot

Trial/Try-Out

Engineers design robots to carry out tasks in locations that are dangerous for people, such as in ocean depths, volcanoes, factories, and war zones. How are robots used during natural disasters (Science and Technical Subjects, RST.6–8.3, RST, 9–10.3)?

Technology

Program and run the robot so that it navigates a maze. Be sure to closely follow the directions provided in the robotic kit (Science and Technical Subjects, RST.6–8.4).

Additional Exercises

1. Use the Internet to locate accounts of robots used in similar types of activities as described. How many projects did you locate? What kinds of robots were you aware of and what new kinds of robots did you learn about (Science and Technical Subjects, RST.6–8.9, RST.9–10.9)?
2. Students can read the book *Maze Runner* and view the movie (optional for students with disabilities). Discuss the creature made from robots in the story.

What can you tell about robotics from reading the story? From the film (Reading and Technology RL 6.7, RL7.7, RL.9–10.7)?
3. Students with disabilities might enjoy the hands-on experience of going to a worksite and observing robots. What did they learn about robotics through observation (Reading and Technology RL 6.7, RL7.7, RL.9–10.7)?

Assessment

For assessment, students will answer the following questions: When programming the robot, what types of problems did you encounter? How did you work through the appropriate measures and steps for the programming of the robot (Science and Technology Subjects RST. 6–8.3, RST. 6–8.1)? (*Handout 6.1A* and *Handout 6.1B*)

Handout 6.1A. The Maze and the Robot

1. Draw a maze on a large sheet.
   a. Step 1: Draw a square.
   b. Step 2: Draw another square inside the first square.
   c. Step 3: Draw more and more squares inside each other.
   d. Step 4: For each square, erase two holes in it near the corners; these will be the entrance and exit holes.
   e. Erase more holes to make the maze more challenging.
   f. Test the route through the maze to make sure there is only one exit.
2. Blindfold a student and provide oral directions to complete the maze. After building the robot from the kit, have the robot go through the maze.

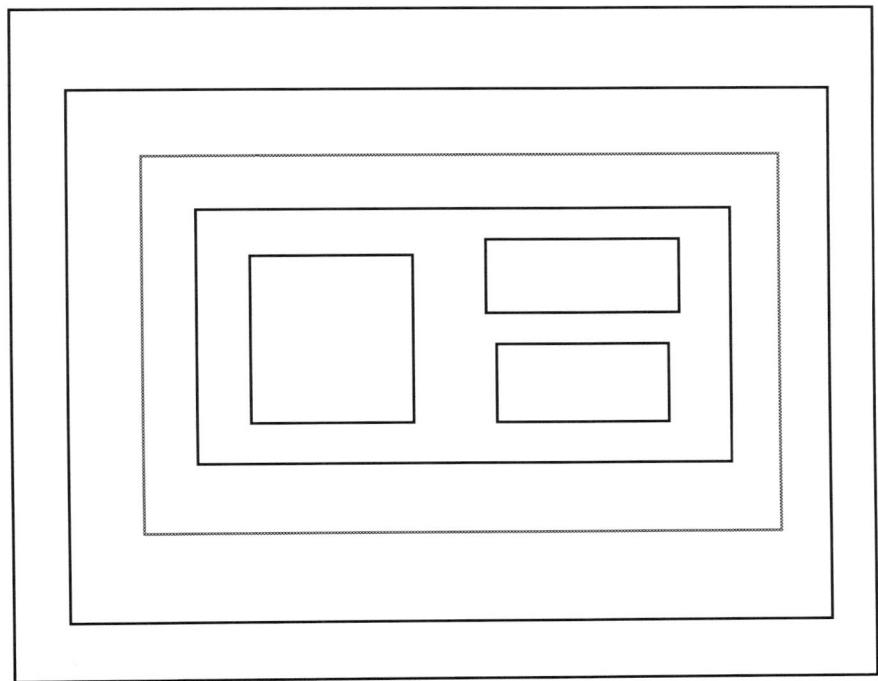

Handout 6.1B. Student Academic Rubric

Category	Below Standard 1	Meet Standard 2	Above Standard 3	Score
Sequencing	Many of the supporting details or arguments are not presented in an expected or logical manner.	Arguments and supporting details are provided in a fairly logical order and are easy to follow.	Arguments and supporting details are provided in logical order using the discussions from the class.	
Paragraph narrative	Only a few arguments are specific to the chart.	Arguments are specific but do not go beyond the class discussion.	Arguments in the paragraph are specific and some even go beyond the class discussion.	
Sentence structure	None of the sentences follow the sequence.	A few sentences are well constructed and follow some sequence.	All the sentences are well constructed and follow the sequence.	
Connection to engineering	The paragraph shows no connection with prior knowledge of engineering.	The paragraph shows some connection with prior knowledge of engineering concepts.	The paragraph shows evidence of thoughtful connections with prior knowledge of the engineering process.	
Comments				

Cognitive Reflection

What are some other creative ways to use robotics other than the ones discussed in this lesson (Science and Technology Subjects RST. 6–8.1, RST.9–10.1)?

Keep, Retain, and Generalize

How will your knowledge of robotics help you in social studies classes (Science and Technology Subjects RST. 6–8.9, RST.9–10.9)?

LESSON 6.2. ENGINEERING CONSTRUCTION AND BUILDING

Harry Potter by J. K. Rowling (1997 to 2016)

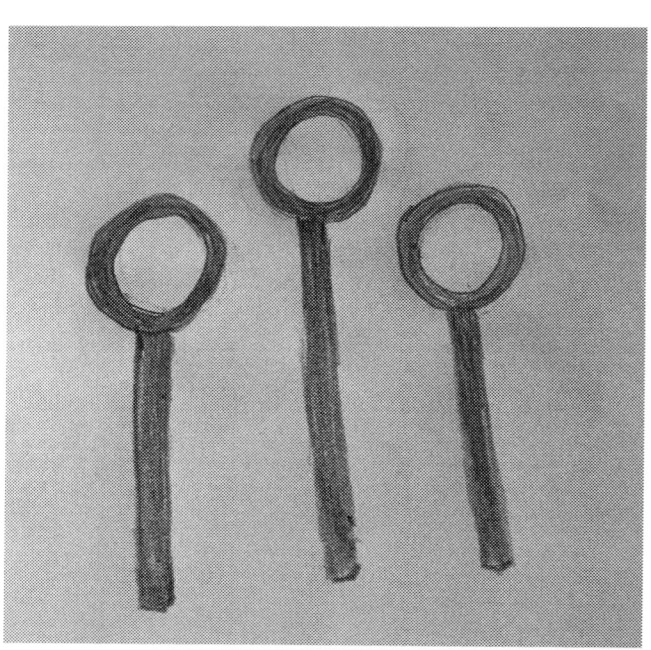

Summary. Harry Potter, the main character, learns on his 11th birthday that he is the orphaned son of two powerful wizards and possesses unique magical powers of his own. He becomes a student at Hogwarts, an English boarding school for wizards. There, he meets several friends and allies who help him discover the truth about his family's shadowy deaths.

Application

Goal. To teach construction skills to adolescent students through examination and the construction of a quidditch stadium. Students will explore chemical, civil, electrical, and mechanical engineering in this lesson.

Objectives

- Students will be able to construct a mini quidditch stadium model.

- Students will be able to connect the skills of several engineering disciplines: civil engineering, architectural engineering, construction engineering, and mechanical engineering.

Standards. To view the Common Core Standards that correspond with this lesson, please visit the *STEAM Meets Story* page on www.tcpress.com and click on the Resources tab.

Teaching Strategy

Introduction

1. Students will watch a short video of the quidditch match from a Harry Potter film, which will provide students with a visual of the venue and architecture they will later model (RST.6–8.3, RST.9–10.3, RST.11–12.3; HSG. CO.D.12) (https://www.youtube.com /watch?v=lwW21jS_bmg).
2. Students will then look at various images of quidditch stadiums to use as references for construction (RST.6–8.3, RST.9–10.3, RST.11–12.3; HSG.CO.D.12).
3. Students will watch a video of a miniature quidditch stadiums built with magnetic balls. Students will also see the construction guides in this video (RST.6–8.3, RST.9–10.3, RST.11–12.3; HSG.CO.D.12) (https://www.youtube.com /watch?v=GCcUNOcTWtM&t=308s).

Materials

- Videos and images
- Computers and/or tablets
- Blueprint and instructions
- Measuring tools and calculators
- Choice of magnet balls, tile magnets, LEGOs, and/ or craft sticks
- Excerpts from Harry Potter books and video clips
- Assessment rubric (*Handout 6.2A*)
- Lesson Reflection (*Handout 6.2B*)

Time. One to two class periods

Essential vocabulary. Students will incorporate key vocabulary when discussing their construction: angle, bisect, congruent, diameter, dimensions, geometric, height, length, measurement, perpendicular, scale, segment, similar, straight line, symmetrical, and width.

Handout 6.2A. Assessment Rubric for the Instructor

Rate the Performance 1 to 5	Score
Utilization of building materials	
Explanation of why they chose a specific material over another to use in construction	
Utilization of videos, images, laptops, and/or tablets for reference	
Plan before constructing	
Teamwork	
Division of work	
How well instructions were followed	
Demonstration of questioning and assessing their progress	
Modifications or innovations to the design	
Replica of a quidditch stadium	
Final product	

Handout 6.2B. Lesson Reflection

Using what you learned about quidditch stadiums, complete the following task:

View the cited blog on construction skills (https://www .bigrentz.com/blog/construction-skills). Select two of the skills that follow. Which helped you the most to build the quidditch stadium? Explain in a paragraph why and how these skills were important for your construction.

- Physical strength and endurance
- Dexterity and hand-eye coordination
- Building and engineering knowledge
- Strong reading and math skills
- Memory
- Communication
- Experience with technology
- Willingness to learn
- Problem solving
- Decisionmaking skills
- Project organization
- Teamwork/collaboration
- Leadership

Trial/Try-Out

1. In groups of five, students will utilize magnet balls, tile magnets, LEGOs, and/or craft sticks to build a model. They will also use calculators and

measuring tools to make precise measurements and assure that the model is true to scale.

2. Students will compare quidditch stadiums to other real stadiums in existence today, such as baseball stadiums, soccer stadiums, football stadiums, and more. They will compare shapes and designs of the fields and structures (RST.6–8.3, RST.9–10.3, RST.11–12.3, HSG.CO.D.12).

3. A quidditch pitch is an oval approximately 500 feet in length × 180 feet in width. However, in quidditch today, the semi-oval or rectangular field is approximately 60 yards (180 feet) in length × 36 yards (108 feet) in width or 66 yards (198 feet) in length × 36 yards (108 feet) in width.(RST.6–8.3, RST.9–10.3, RST.11–12.3, HSG.CO.D.12). Students will select the pitch they prefer, semi-circle or rectangular. They will recall and refer to the image/field of choice and then go to the school yard (whether grass or asphalt) and measure its length and width. They will determine if the standard layouts already fit or if their dimensions need to be reconfigured/scaled down by using proper math formulas and measurements. Once ascertained, students will provide the right measurements. They will then create their own blueprint and outline the school yard using either chalk, paint, tape, cones, or other tools. Students will then go to the school's gym and determine the proper scaled-down dimensions to play quidditch in that space. Once discovered, they will create a second blueprint. They will measure the perimeter and place tape or cones to outline the space (RST.6–8.3, RST.9–10.3, RST.11–12.3; HSG.CO.D.12).

Assessment

Students will be evaluated based on the rubric in *Handouts 6.1A* and *B* (RST.6–8.3, RST.9–10.3, RST.11–12.3; HSG.CO.D.12).

Cognitive Reflection

Stadiums, residential buildings, airports, and other structures around the world are constructed by multiple people who collaborate with each other. Why is it important in construction engineering for various people to work together and for each one to understand measurements (RST.6–8.3, RST.9–10.3, RST.11–12.3, HSG.CO.D.12)?

Keep, Retain, and Generalize

1. Before construction of a stadium, residential building, subway, airport, or other structure, many architecture and engineering firms construct miniature models, in addition to designing images and blueprints. After having built your own miniature quidditch stadium, why do you think it is important for architects and engineers to build miniature models (RST.6–8.3, RST.9–10.3, RST.11–12.3, HSG.CO.D.12)?

2. After this lesson, describe in a paragraph how a quidditch stadium is connected to one of the following categories of engineering: civil engineering, architectural engineering, construction engineering, or mechanical engineering. Provide examples to support your response.

LESSON 6.3. DRONE ENGINEERING

Men in Black by Steve Perry (1994)

Summary. Men in Black is a top-secret agency committed to pursuing and monitoring the activities and actions of aliens. James Edwards is a streetwise policeman recruited by Agent K to become Agent J. While the world ponders and watches for signs of alien civilizations, these men know that alien beings are presently walking among us in human form. These men fight to preserve the peace between the aliens and humans while remaining undetected.

Application

Goal. In one of the three *Men in Black* movies, Agent J fights several drones. Students will learn about drones.

Objectives

- Students will know several uses of drones and the rudiments for constructing a drone.

Standards. To view the Common Core Standards that correspond with this lesson, please visit the *STEAM Meets Story* page on www.tcpress.com and click on the Resources tab.

Teaching Strategy

Introduction. Agent K is battling with an alien who successfully uses drones in *Men in Black II* (https://www.youtube.com/watch?v=5NY_8ulSutc). Review the definition of a drone (https://www.youtube.com/watch?v=zfKkrnsJtFw) and the top 10 ways drones are used (https://www.youtube.com/watch?v=3MU2vcKCL5s). Explain what drones are and name, demonstrate, and describe some important ways they are used in society (RST.6–8.3, RST.9–10.3, RST.11–12.3, HSG.CO.D.12).

What is drone racing? How are drones used to control traffic congestion (RST.6–8.3, RST.9–10.3, RST.11–12.3, HSG.CO.D.12)?

Materials

- Videos and images
- Computers and/or tablets
- An already constructed drone to replicate
- Blueprint and instructions
- Construction pieces from the kit, as seen in the videos and images

Time. One to two class periods

Essential vocabulary. Students will incorporate key vocabulary when discussing drones. (See *Handout 6.3A*.)

Trial/Try-Out

1. Students will construct a drone after watching the two video guides (RST.6–8.3, RST.9–10.3, RST.11–12.3, HSG.CO.D.12).
   a. DIY Building Blocks Drone STEM Kit (https://www.youtube.com/watch?v=_EvC2lVVFDo)
   b. TR-D5 DIY Drone by Top Race! (https://www.youtube.com/watch?v=jZrjUDsBxbA)
2. Students will work in a team to build small drones, utilizing the pieces from a DYI kit. Once complete, students will have a drone race. Students can play a version of quidditch with their drones. This will give students another vantage point on how flying in a quidditch game looks and how they can maneuver characters, as they would a drone, by

Handout 6.3A. Basic Drone Vocabulary

The use of drones, also called unmanned aircraft systems, whether used for recreation or for business, is a highly regulated activity in the United States. Find the definition of 10 of the words you believe might be useful.

Word	Definition
Above ground level (AGL)	
Anemometer	
Controller	
Drone	
Flight controller	
First-person view (FPV)	
Flyaway	
FSDO	
Geofencing	
Gimbal	
GIS	
Global navigation satellite system (GNSS)	
Global positioning system (GPS)	

using a controller. Students can also design their own racing system. A LEGO toy, representing a character, can also be attached on top of the drone (RST.6–8.3, RST.9–10.3, RST.11–12.3, HSG.CO.D.12).

Technology. Students can build drones with on-board cameras and pilot those using smartphones or a controller.

Additional exercises. Students will compare the drones they've built to other piloted machines such as airplanes and helicopters (RST.6–8.3, RST.9–10.3, RST.11–12.3, HSG.CO.D.12).

Assessment

Students will be evaluated on a scale of 1–5 on the workability of the drone they built.

Cognitive Reflection

1. Why do you think it is important for engineers to build drones, and how can drones help us to see things that we couldn't before (RST.6–8.3, RST.9–10.3, RST.11–12.3, HSG.CO.D.12)?
2. *Men in Black* displayed a number of innovative technological advancements. Which ones do you think will be realized in the next decade? In the next millennium (RST.6–8.3, RST.9–10.3, RST.11–12.3)?

Keep, Retain, and Generalize

1. Today, drones are being used and conceived for numerous purposes. These include surveying areas, construction, security, telecommunications, forecasting, animal conservation, agriculture, food and product delivery, medicine and supplies transportation, photography, filming movies, and human transportation, among many others. Many engineers are looking at different ways to utilize drone technology. After having built your own drone, in what other new, productive ways do you think drones can be used to benefit society in the future (RST.6–8.3, RST.9–10.3, RST.11–12.3, HSG.CO.D.12)?
2. There was a lot of fighting for protection of the Earth in *Men in Black*. Do you think that drone technology can be used in battles or wars? How? What can you find on the Internet (RST.6–8.3, RST.9–10.3, RST.11–12.3, HSG.CO.D.12)?

LESSON 6.4. ENGINEERING TRAINS

Snowpiercer by Jacques Lob (2014)

Summary. Survivors of Earth's second ice age live out their days on a luxury train that ploughs through snow and ice. The train's poorest residents, who live in the squalid caboose, plan to improve their lot by taking over the engine room. The storyline begins with an attempt to counteract global warming by engineering climate patterns with a chemical called CW7. The plan backfires and causes an ice age, which kills all life on Earth except for the people who live on the Snowpiercer, a perpetual motion train that travels a span of track that loops one time around the globe in a year. It was created by a business magnate named Wilford, who runs the train. The train has 101 cars; and the rich live in the front in luxury and the poor live in the rear. One of the citizens of the rear plans a revolt and the revolt begins.

Application

Goal. Students will learn about trains.

Objectives

- Students will be able to construct a model of a magnetic levitation train and an electromagnetic train.
- Students will expand their construction and innovation skills as well as make connections to construction engineering, transportation engineering, railway/train engineering, and electromechanical engineering.

Standards. To view the Common Core Standards that correspond with this lesson, please visit the *STEAM Meets Story* page on www.tcpress.com and click on the Resources tab.

Teaching Strategy

Introduction. Students will begin the lesson by viewing clips from the Snowpiercer movie (https://www.youtube.com/watch?v=l8zzr5XrF_w). Students will then learn that high-speed trains, like maglevs, are used today around the world to transport people at greater speeds than traditional trains. There are two types of designs and technology used for maglevs. These are electromagnetic suspension (EMS) and electrodynamic suspension (EDS). To further understand this, students will watch videos on maglevs and how they work. Here is a video of the way these trains can revolutionize travel (https://www.youtube.com/watch?v=Wor8C3ZIAu8). These trains use powerful magnets to work (https://www.youtube.com/watch? v=m-rNILcfTKM).

In groups, students will be instructed to build a maglev. KiwiCo (https://www.youtube.com/watch?v=Fd3O6sXBXqA) provides an example. There are others on the Internet students can also can try (RST.6–8.3, RST.9–10.3, RST.11–12.3, HSG.CO.D.12).

Materials

- Videos
- Computers and/or tablets
- Blueprint and instructions
- Measuring tools and calculators
- Tools and supplies from the videos
- Handouts
- Excerpts from *Snowpiercer*

Time. One to two class periods

Essential vocabulary. Students will incorporate key vocabulary when discussing their construction. These vocabulary words and an exercise to accompany them are listed in *Handout 6.4A*.

Trial/Try-Out

1. Students will read an article about magnetic trains (https://www.physics-and-radio-electronics.com/blog/magnetictrain-maglevtrain/). Then, they will combine their tracks as a class to build a large racing track, similar to a electromagnetic track, where they can race their trains (RST.6–8.3, RST.9–10.3, RST.11–12.3, HSG.CO.D.12).
2. As a group, students will use Internet resources to compare the shapes, designs, speed, cost, and utility of maglev trains to traditional trains on rails. They will also compare maglevs to planes and other vehicles (https://www.physics-and-radio-electronics.com/blog/magnetictrain-maglevtrain/) (RST.6–8.3, RST.9–10.3, RST.11–12.3, HSG.CO.D.12).
3. Students will research online where maglev and high-speed trains are located around the world.

They will inquire as to why maglevs aren't used in the United States yet for interstate travel, what types of trains are currently used, and how many high-speed trains exist. They will also investigate current plans for future high-speed trains in the United States (RST.6–8.3, RST.9–10.3, RST.11–12.3, HSG.CO.D.12).

Assessment

Students will be evaluated based on a rubric. How well did they construct the train? Were they able to remember some of the train-related vocabulary and use the vocabulary in the assignments provided (RST.6–8.3, RST.9–10.3, RST.11–12.3, HSG.CO.D.12)?

Cognitive Reflection

1. In the *Snowpiercer* movie and graphic novel, the train has a caste system. There are the very rich in

Handout 6.4A. The ABCs of Trains

There are words that begin with A, B, or C. Please provide the definition to at least 20. Put a check next to the word that might be in your textbook.

Words	Definition	Check
Adhesion railway		
Adhesive weight		
Air brake		
Alerter or watchdog		
All weather adhesion		
Alternator		
The American wheel arrangement		
Angle cock		
Articulated locomotive		
Articulation		
Ashpan		
Asynchronous		
Atlantic type		
Automatic block signaling (ABS)		
Automatic train control (ATC)		
Automatic train operation (ATO)		

(continued)

Words	Definition	Check	Words	Definition	Check
Automatic train protection (ATP)			Builder's plate		
Auto train			Bulkhead flatcar		
Axle box			Bungalow		
A Swiss axle box			Bustitution		
Back head			Cab		
Bail off			Cab forward		
Balancing			Cab-less		
Balise			Caboose		
Ballast			Cant		
Ballast tamper			Car body unit		
Balloon			Catenary		
Bay platform			Center beam		
Beep			Chord		
Bellmouth			Co-Co locomotives (EU)		
Berkshire type			Coal pusher		
Blastpipe			Color light signal		
Block section			Color position signal		
Bo-Bo (Europe)			Combined power handle		
Bogie			Compound locomotive		
Boiler			Configurable system		
Bolster			Consolidation wheel arrangement		
Boom barrier			Container on flat car (COFC)		
Boom barriers at a railway crossing in France			Continuous welded rail (CWR)		
A pivoted road barrier at a level crossing			Control car		
Booster engine			Control system		
Brakeman's cabin or brakeman's cab			Coupling rods		
Brake pipe			Covered goods wagon		
Branch line			Cow-calf or cow and calf		
Brick arch			Crank pin		
British Rail Universal Trolley Equipment (BRUTE)			Crosshead		
			Cut		
Broad gauge			Cut lever		
Bubble car			Cut off		
Buckeye coupler			Cutting		
Buffer			Cycle braking		
Buffer stop or bumper post			Cylinder		
			Cylinder cock		

the front of the train, the working class in the middle, and the poor at the rear. The very rich have the best, the middle class keep the train running and serve as the workers, and the poor receive the scraps and live in the worst conditions. How does the train mirror society? Is this the way the United States is arranged? What about other countries? What systems might they have? List three countries with various systems. Explain how their systems work in *Handout 6.4B* (L.6.4.c, L.7.4.c, L.8.4.c, L.9–10.4.c, L. 6.4.d/ L.7.4.d L.8.4.d/L.9–10.4.d, WHST 6–8.2.d, WHST 9–10.2.d).

Handout 6.4B. Same or Different

Name three countries and compare how the systems of government operate the same and differently.

	Country 1	Country 2	Country 3
Similar operations			
Different operations			

2. Regarding seating on the train, are there differences in seating? How are they similar to or different from what happens in *Snowpiercer* (L.6.4.c, L.7.4.c, L.8.4.c, L.9–10.4.c, L. 6.4.d/ L.7.4.d L.8.4.d/L.9–10.4.d, WHST 6–8.2.d, WHST 9–10.2.d)?

Keep, Retain, and Generalize

High-speed trains are expanding across Europe, China, Japan, and South Korea. Presently, engineers in the United States are diligently seeking ways to covert the national interstate rail system into high-speed rail in a vast effort to link cities in shorter commuter times. After having built your own maglev and electromagnetic train, learning about this technology, as well as learning about high-speed rail, what valid reasons would you give for the United States to invest in high-speed trains in the immediate future? How could high-speed trains benefit the United States and its citizens? Which major cities should be prioritized for high-speed rail connection (RST.6–8.3, RST.9–10.3, RST.11–12.3, HSG.CO.D.12)?

LESSON 6.5. SATELLITE ENGINEERING

Star Trek—The Motion Picture: A Novel by Gene Rodenberry (1979)

Summary. The original 5-year mission of the Enterprise has ended. James T. Kirk is now an admiral. All of the Starship Enterprise's original crew have other jobs, but they have all been contacted for one last mission. The

Starship Enterprise is new and now has a mission to find the power that threatens all mankind on the Earth. An immense cloud has appeared that can collide with the Earth. After researching, it is noted that the cloud contains artificial intelligence (AI) that is determined to kill all the "carbon units affecting earth." A crew member is "kidnapped" and is thereby turned into a probe. She becomes an android lock-alike containing her memories and comes back representing Vger. Later we find that the Voyager 6 (Vger) satellite sent from Earth to gather data has come home to meet the creator.

Application

Goal. The goal of this lesson is to focus on satellites and satellite engineers.

Objectives

- Students will be able to name the purpose of satellites.

- Students will be able to identify the role of a satellite engineer.
- Students will be able to identify other jobs a satellite engineer does.

Standards. To view the Common Core Standards that correspond with this lesson, please visit the *STEAM Meets Story* page on www.tcpress.com and click on the Resources tab.

Teaching Strategy

Students will view the movie clip that reveals Voyager 6 (https://www.youtube.com/watch?v=US-DF12DpVk). NASA space operations collect technical data to help engineers gather knowledge about Earth's environment, other planets, the solar system, and the universe. These missions direct future exploration. There are three types of robotic explorers: (a) satellites, (b) landers and (c) rovers. Each type of robotic explorer has its merits and is used for different purposes depending on the planet being observed and the types of observations desired. What is the difference in the three? What do engineers have to do when they build a satellite (https://www.youtube.com/watch?v=Ti-4sB3mEb8)? NASA (https://science.nasa.gov/astrophysics/universe-spacecraft-paper-models) provides a model (RST.9–10.4. RST.9–10.6, RST.9–10.3).

Materials

- Glue or tape
- Styrofoam
- Pen or marker
- Scissors or knife with ruler and cutting board
- Styrofoam block: 3 cm wide × 6 cm deep × 6.3 cm deep
- Thick double-sided sticky tape (optional)
- Four toothpicks (optional)
- Two wooden barbeque skewers: 20 cm
- Copies of books, excerpts from the book, the movie, or selected clips from the movie

Time. Three class periods

Essential vocabulary. Define the words on *Handout 6.5A*. These words are valuable if you are learning about satellites (RST.9–10.10, RST.9–10.5).

Trial/Try Outs

1. What is a satellite? Ask students what they know about satellites and their usage.

Handout 6.5A. Essential Vocabulary

Provide definition a for 20 words.

Debris	
Astronaut	
Asteroid	
Atmosphere	
Cosmos	
Earth	
Explorer	
Galaxy	
Gas	
Horizon	
Launch	
Meteor	
Moon	
Ocean	
Orbit	
Outer space	
Planet	
Radiation	
Rocket	
Simulator	
Space	
Spacecraft	
Space shuttle	
Space station	
Surface	
Universe	
Weightlessness	
Commercial	
Cosmic	
Extreme	
Gravitation	
Gravitational	
Horizontal	
Inevitable	

Lunar	
Meteoric	
Outer	
Solar	
Terrestrial	
Toxic	
Uninhabitable	
Universal	
Unmanned	
Lacking a crew	
Acclimate	
Colonize	
Explore	
Float	
Orbit	
Propel	
Rotate	
Sustain	
Simulate	
Undergo	

2. Ask students to name different satellites. Are there natural satellites? What are they? What is the difference in those and manmade satellites (RST.9–10.6)?
3. Satellite engineers develop, design, and repair various satellites related to technology. Other responsibilities include installing, testing, and redesigning satellites and related technology. They also create quality standards for manufacturing satellites and satellite operations. What engineers perform similar jobs (RST.9–1-.10)?

Assessment

Explain the main parts of a human-made satellite and point out examples: main bus, power system, science instruments, and communication antennas. Drawing a diagram might be helpful. You can form a team of two and search the Internet for assistance (RST.9–10.7).

Cognitive Reflection

Captain Kirk speaks about Voyager 6 attaining AI in the book and the movie. He explains Voyager 6 was in need of repair and another computer joined with it, repaired it, and the satellite achieved AI. Any device that can recognize its environment has cognitive functions, can problem solve, and can take actions to attain its goals can have AI. Do you believe machines can achieve or will attain AI in the future such as Vger did? Search the Internet to see if you can find evidence in support of or against this claim. Write debate points for each side (RST.9–10.9).

Keep, Retain, and Generalize

Watch Learn Engineering's video (https://www.youtube.com/watch?v=ror4P1UAv_g) to learn more about satellites. Many people believe that members of CLD populations have difficulty obtaining engineering jobs. Can you find statistics on this opinion? What can be done to increase the number of CLD students in these types of fields?

LESSON 6.6. ENGINEERING CYBORGS

The Terminator by James Cameron (Director/Screenwriter) and Gale Anna Hurd (Producer/Screenwriter) (1984)

Summary. A cyborg assassin known as the Terminator travels from 2029 to 1984 to kill Sarah Connor. Kyle Reese is sent by Sarah's son to protect her. He tells her about Skynet's AI. Sarah is selected because her unborn son will lead the fight against them. The strong, determined Terminator stops at nothing while Kyle and Sarah attempt to escape. Sarah and Kyle fall in love as the Terminator peruses them relentlessly, killing a number of her relatives and friends along the way. Finally, the badly damaged Terminator chases them into a factory where he kills Kyle; Sarah lures Terminator into a hydraulic press and crushes it. Later you find that Sarah is pregnant and is recording audio tapes for her unborn son, John. A boy takes an instant photograph of her, and she buys it; it's the same photograph that John will give to Kyle. There have been a series of *Terminator* movies and books, all of them revolving around efforts to stop Skynet.

Application

Goal. Students will understand the cyborgs and accompanying technologies.

Objectives

- Students will be able to name five technologies related to cyborgs.
- Students will be able to name professions related to cyborgs.

Standards. To view the Common Core Standards that correspond with this lesson, please visit the *STEAM Meets Story* page on www.tcpress.com and click on the Resources tab.

Teaching Strategy

Science and engineering work in tandem when creating cyborgs. A clip from *Terminator* shows the what the cyborg was built from (https://www.youtube.com /watch?v=k4gODyZKE1Q). The engineers make the mechanical and the technological parts and the

doctors implant these machines in the person. These implants within the person create a relationship between the flesh and the machine. What is a cyborg? Students will determine the difference between real cyborgs and science fiction cyborgs. What types of engineers work on cyborgs? What did you find in your research that has to do with advanced robotics and tissue combinations (MP4, MP7)?

Materials

- The movie or selected clips from the movie

Time. Two class periods

Essential vocabulary. Define the words in *Handout 6.6A*. These words are valuable if you are learning about cyborgs (RST.9–10.4).

Trial/Try Outs

In the *Terminator* series, all of the machines are sent back from a future where cyborgs and robots ruled the Earth. Read the article "Six Ways Technology is Turning Us into Cyborgs" (https://medium.com/lassondeschool/6-ways -technology -is-turning-us-into-cyborgs-c9262d78d886). Provide a summary of the article. Is this a good thing or a bad thing? Provide your opinion and then search the Internet and find information to support your opinion (RST.9–10.1, RST.9–10.7).

Handout 6.6A. Cyborg Definitions

Cybernetics	
Bionic	
Organism	
Cylon	
Robot	
Cyberpunk	
Android	
Cybermen	
Automaton	
Transhumanism	
Doppelganger	
Clone	

Assessment

This assignment will be assessed with the completion and review of the article. Students will receive points from 1 to 5 depending on the number of cyborg technologies they discover from the article and how many they find from their research. Ten technologies and their explanations will receive a 5, eight technologies will receive a 4, six technologies will receive a 3, four a 2, and one a 1.

Cognitive Reflection

Cyborgs were used in the *Star Wars* franchise. For example, Darth Vader and Luke Skywalker had mechanical parts. In these books and movies, Anakin Skywalker becomes Darth Vader when he pledges to become a Sith lord rather than a Jedi knight. He then becomes badly burned while in a fight with Obi Wan Kinobi. Afterward, he becomes a cyborg made of machine and flesh. In the famous "Luke I am your father" scene, Luke loses his hand (https://www.youtube.com/watch?v=Lbjru5CQIW4). Later, Luke gets a cyborg hand (https://www.youtube.com/watch?v=_PNU84bbass). Do you think cyborgs like this can be created in the future? What cyborgs exist presently (RST.9–10.6)?

Keep, Retain, and Generalize

Even *Star Trek* had a cyborg species, the Borg. These were a conquering species that attacked other species and consumed them in the Borg hive mind (the Borg queen descends from the ceiling in first contact; https://www.youtube.com/watch?v=jibEwOHr2t8). Think of a disability a person may have and what kind of cybernetic organism could help that person. Conduct research to see if it already exists. Research with a team of four and find out what kinds of technologies can help persons with disabilities. (Watch the film about a person with a cybernetic eye; https://youtube/JWMYW-SkURI.) Many of these devices are costly and many persons with disabilities may not be able to afford this technology. What could we do in our society to address that (MP4, MP7)?

LESSON 6.7. SOFTWARE ENGINEERING

Jumanji by Chris Van Allsburg (1984)

Summary. There were four installments to the *Jumanji* franchise. The franchise includes four films: *Jumanji* (1995), *Zathura: A Space Adventure* (2005), *Jumanji: Welcome to the Jungle* (2017), and *Jumanji: The Next Level* (2019); as well as an animated television series that aired from 1996 to 1999. *Jumanji: Welcome to the Jungle* is about four high school kids who discover an old video game console and are sucked into the setting and become part of the game's software. They become avatars in the game. They must play and survive to get out of the software or they'll be stuck forever. To beat the game and return to the real world, they'll have to go on the most dangerous adventure of their lives. They discover that Alan Parrish left 20 years ago and is left in the game from the 1995 adventure.

Application

Goal. Students will understand what a software engineer does.

Objectives

- Students will be able to name five things a software engineer does.
- Students will be able to name three recent innovations of software engineers.

Standards. To view the Common Core Standards that correspond with this lesson, please visit the *STEAM Meets Story* page on www.tcpress.com and click on the Resources tab.

Teaching Strategy

Software engineers develop systems and software for businesses. These products range from business applications and games to network control systems and operating systems. A software engineer's responsibilities may also include the following:

- Determining software needs
- Designing, developing, and testing a system applications
- Drawing diagrams and models that help developers create code
- Recommending software upgrades
- Collaborating with developers and other engineers to create software

Read and summarize Ollason's article (https://www.freecodecamp.org/news/lessons-from-2-5-years-of-software-engineering-da66891f1b09/) about a software engineer who began working at age 16 designing software (MP4, MP5, MP6).

Materials

- Copies of books, excerpts from the book, the movie, or selected clips from the movie

Time. Two class periods

Essential vocabulary. Define the words in Handout 6.7A. These words are valuable if you are learning about cyborgs (RST.9–10.7, RST.9–10.4).

Trial/Try Outs

This clip of *Jumanji* describes the avatars in the game (https://www.youtube.com/watch?v=dSKIEjS3f9Y). If you pay any attention to the technology world, you've undoubtedly heard the terms *coding* and *programming* dozens of times. Many of the best tech careers require the ability to code. If you want to work in a high-paying field like software engineering, web development, or data science, understanding and using code is essential. Research and answer the following questions: What is coding? What can programming be used for? You can read Bit Degree's article (https://www.bitdegree.org/tutorials/what-is-coding/) for answers (RST.9–10.2, RST.9–10.6, RST.9–10.8).

Assessment

List the careers that a software engineer can have.

Handout 6.7A. Software Engineer Definitions

Abstraction	
Accessibility	
Algorithm	
Binary	
Binary alphabet	
Bit	
Block-based programming language	
Blockly	
Bug	
Byte	
Call (a variable)	
Call (a function)	
Click	
Code	
Command	
Computational thinking	
Computer science	
Conditionals	
Crowdsourcing	
Cyberbullying	
Data	
Debugging	
Decompose	
define (a function)	
digital citizen	
digital footprint	
DNS (domain name service)	
Double-click	
Drag	
Drop	
DSL cable	
Event	
Event handler	
Fiber optic cable	
For loop	

Frustrated		Tool box	
Function		Trustworthy	
Function call		Try	
Function definition		URL (universal resource locator)	
If statement		Username	
Input		Variable	
Internet		Website	
IP address		While loop	
Iteration		Wi-Fi	
Loop		Workspace	
Online			
Output			
Packets			
Pattern matching			
Parameter			
Pixel			
Program			
Programming			
Repeat			
Run program			
Search engine			
Servers			

Cognitive Reflection

There are a number of websites where you can build your own app. In a team, build an app. You can use the website Flutter or research your own (RST.9–10.3).

Keep, Retain, and Generalize

Computer coding is the use of computer programming languages to give computers and machines instructions on what actions to perform. It is the way humans communicate with machines, and it allows us to create software like programs, operating systems, and mobile apps. Go to the code.org and perform an hour of code. Choose your grade level (MP2).

LESSON 6.8. AEROSPACE ENGINEERING

Minority Report by Phillip Dick (1956)

Summary. *Minority Report* is an action-detective thriller set in Washington, DC in 2054, where police used specialized mutated humans, called precogs, to predict murders before they happen, reducing the murder rate to zero. The crime unit uses specialized software and helicopter-like technology to stop the crime after it is predicted. Premurderers are imprisoned in virtual reality and the government is on the verge of adopting the controversial program nationwide. A strong theme of the film is free will versus determinism and making arrests and jailing individuals before a crime is committed. Other themes include the role of the government attempting to protect the population. The book and the film are full of technological advancements that promote engineering.

Application

Goal. Explore aerospace engineering

Objectives

- Students will be able to name at least five duties of aerospace engineers.
- Students will be able to name two principles related aerodynamics with 80% accuracy.

- Students will be able to name five principles with 80% accuracy.

Standards. To view the Common Core Standards that correspond with this lesson, please visit the *STEAM Meets Story* page on www.tcpress.com and click on the Resources tab.

Teaching Strategy

Aerospace engineers specialize in astronautics or aeronautical engineering. They apply science and technology inside and outside the Earth's atmosphere. They study the aerodynamic performance of aircraft and propulsion systems in aircraft designs. Aerodynamics involves the forces of air as it interacts with solid objects such as airplane wings. Helicopters use the principles of aerodynamics to move through the air. Airplanes are able to fly because of two key principles: first, the "thrust" given by the airplane engines propels it through the air; second, the movement of air over the wings produces the lifting force needed to keep the plane in the air. Watch Engineering Videos' (https://www.sciencekids.co.nz/videos/engineering/flightaerodynamics.html) film about aerodynamics and watch the scene in *Minority Report* that provides the gist of what the Precrime Unit does (https://www.youtube.com/watch?v=BmSarhudhiY). There is a lot of technology in the scene. You see the helicopter plane the unit uses. Does it operate by the same lift and thrust principal (MP1, MP4, MP5, MP6, MP7, MP8)?

Materials

- Copies of books, excerpts from the book, the movie, or selected clips from the movie
- Water bottle (chassis)
- Balloon, vinyl tubing, rubber band (motor)
- Wooden skewers and straws (axle)
- Various materials for wheels
- Tape

Time. Two class periods.

Handout 6.8A. Vocabulary

Bernoulli's principle	
Thrust	
Gravity	
Drag	
Lift	
Upward force	
Trajectory	
Aerospace	
Anemometer	
Orbit	
Latitude	
Longitude	
Propulsion	
Stabilizer	
Rudder	
Jet stream	
Altimeter	
FAA	
Federal Aviation Administration	
Fuselage	
Elevator	
Doppler effect	
Barometer	
Calibration	
Atmosphere	
Altitude	

Essential vocabulary. Define the words in *Handout 6.8A*. These words are valuable if you are learning about aerospace engineering (RST.9–10.3, RST.9–10.5).

Trial/Try Outs

"It's not rocket science!" Why sure it is. Who does this kind of work? Aerospace engineers do! Aerospace engineering is work with aircraft and related systems operating in the Earth's atmosphere (aeronautical engineering) as well spacecraft missiles, rocket-propulsion systems, and other equipment operating beyond the Earth's atmosphere (aeronautical engineering). What other specific duties do they have?

Assessment

The thrust of a jet engine is the same as that of a balloon when the air is allowed to escape. The air pushes on the balloon so that it stays inflated. Covering the end of the balloon maintains high pressure in the balloon. When the balloon is opened, the pressure forces the balloon to move. In a jet engine the air enters the engine; it is then heated and compressed and propels the engine from the pressure that is created and acceleration occurs. The amount of acceleration depends on the force and the mass of the object. This principle follows Newton's second law of motion: force = mass × acceleration. Building a balloon car will demonstrate this principle (https://www.youtube.com/watch?v=dR2C1GGJ-9o). You need a water bottle, balloon, vinyl tubing, rubber band, wooden skewers, straws, wheels, and tape. In a team, someone should blow the balloon up a couple times to stretch it out. Cut the straw into two pieces equal to the width of the water bottle. Attach the two straw pieces underneath the water bottle where you feel the front and rear wheels should go. Keep the straws lined up so the car travels in a straight line. Cut two pieces of the wooden skewer. Put one end of each wooden skewer through your wheel. Slide the skewers through the straw and attach the rest of the wheels to the skewer. Then insert the nozzle partway into the balloon. Use a rubber band and secure the nozzle to the balloon. Insert the nozzle through the slit on the top of your water bottle.

Make sure about an inch of the nozzle is sticking out of the mouth of the bottle. Blow up the balloon and pinch the balloon at the base so the air won't escape. Line the rocket car up on the starting line and when the track is clear release the balloon (https://www.grc.nasa.gov/www/k-12/BGP/Ashlie/BalloonRocketCar_easy.html). Watch it go! Explain what the team has done and why this represents a jet engine. See *Handout 6.8B* for the grading rubric (RST.9–10.3).

Handout 6.8B. Balloon Activity Rubric

Name of team: _____

1. Necessary parts were present on the model constructed.

Poor	Average	Good	Excellent
1 2	3 4 5	6 7	8 9 10

2. All parts of the engine materials were used.

Poor	Average	Good	Excellent
1 2	3 4 5	6 7	8 9 10

3. Team collaborated well.

Poor	Average	Good	Excellent
1 2	3 4 5	6 7	8 9 10

4. Team could explain how engines work in a clear and correct manner.

Poor	Average	Good	Excellent
1 2	3 4 5	6 7	8 9 10

5. Team stayed on task.

Poor	Average	Good	Excellent
1 2	3 4 5	6 7	8 9 10

Total: _____

Cognitive Reflection

Thrust moves any aircrafts created by the propulsion system of the aircraft. There are many different propulsion systems that add to the thrust of the plane in many different ways, depending on the type of engine. Thrust is generated by Newton's law: For every action there is an equal and opposite reaction. Fluid (gas) is used in these systems and thrust generated depends on the mass flow through the engine and the exit velocity of the gas. Since there are various kinds of aircrafts, search the Internet to find various types. List at least four. Which is the most powerful? Which is least powerful? Provide pictures and list some rudimentary parts, including air, thrust, lift, drag, and weight (MP1, MP4, MP5, MP6, MP7, MP8).

Keep, Retain, and Generalize

In the book and movie *Minority Report*, the precogs were detained to do the work and were a main part of the system. What about the rights of these pre-cogs?

Additionally, people were arrested before they did the crime. Could there have been another outcome? This system had limited crime to zero. What about imprisoning people who have not yet committed a crime?

Write what you think about this system. What changes would you make to the Precrime Unit (RST.9–10.9)?

LESSON 6.9. ENGINEERING AND SPACE TRAVEL

The Martian by Andy Weir (2011)

Summary. In 2035, NASA sends a man to Mars for 1 month. A large dust and windstorm appears and threatens the transportation vehicle that was provided for mobility on the planet. There is an evacuation attempt, the mode of communication is disturbed, and the astronaut is not able to contact Earth. Mark Watney is a botanist and engineer who uses his skills to stay alive on the planet. He has a minor injury and must rely on his own resourcefulness to survive. He begins a log of his experiences. He begins growing food and creates a habitat. NASA finally discovers that he is alive. NASA and a team of international scientists work tirelessly to bring "the Martian" home, while his crewmates concurrently plot a daring, if not impossible, rescue mission. As these stories of incredible bravery unfold, the world comes together to root for Watney's safe return. A NASA team is dispatched to rescue Mark.

Application

Goal. Students will know the types of engineers who do space travel and what they need to do.

Objectives. Students will explore the following:

- Why people want to visit other planets or travel in space
- What types of engineers are interested in space travel and space rockets and technology
- What endeavors may be in the future for space travel

Standards. To view the Common Core Standards that correspond with this lesson, please visit the *STEAM Meets Story* page on www.tcpress.com and click on the Resources tab.

Teaching Strategy

In the movie and book *The Martian*, no human life is presented on other planets. But the main character does show how he survived by growing food (https://www.youtube.com/watch?v=BH-UmA5Lt3g). Do you think life exists on other planets? How could we find out?

Because he learned to grow food on Mars, do you think we could someday colonize other planets in our solar system? Why would we want to? What would we need to figure out to make that possible (MP1, RST.9–10.2)?

Materials

- Copies of books, excerpts from the book, the movie, or selected clips from the movie

Time. Two class periods

Essential vocabulary

Chemical reaction: A process whereby one type of substance is chemically converted to another substance involving an exchange of energy
Gravity: The natural force of attraction between any two massive bodies
Rocket: A vehicle that moves by ejecting fuel
Star: A huge burning sphere of gas, made up of roughly 90% hydrogen and 10% helium
Thrust: The forward-directed force on a rocket in reaction to the ejection of fuel

Trial/Try Outs

1. Astronauts live in the International Space Station (ISS), and they operate much like the astronaut in *The Maritain*. They perform experiments, use equipment, go up in rockets, and perform other activities that can only be done in space. The "Martian" could survive because he had studied living in this type of environment. These astronauts have to figure out what to do for an extended period of time if they indeed have to colonize other places in the solar system. The environment they live in is much like the movie and book. It is about the size of a football field with different compartments. This clip shows you how they

live (https://www.nasa.gov/multimedia/nasatv
/iss_ustream.html). Can you write a description
of what you observed (RST.9–10.3)?

2. Space engineers usually work at NASA and design
and construct spaceships. They also work for
private companies that privatize space travel and
similar advances in aerospace technology. Space
engineers use math and science to create new
technology. They improve existing technology for
both airplane flight and space flight. Have students
research how the ISS is built, and what it would be
like to live on it. In teams, get students to consider
how scientists and engineers build the ISS and get
it to successfully orbit the Earth. What kinds of
things could affect the orbit (MP1, MP4, MP5,
MP6, MP7, RST.9–10.8)?

Assessment

A lunar base or space station is a self-contained habitat necessary for the survival of animals and plants.
In groups of four, students will research and make a
written or 3D model of one of the nine life support
systems (air supply, communications, electricity,
food production and delivery, recreation, temperature control, transportation, waste management, and
water supply). Prepare a 10-minute presentation
(*Handout 6.9A*).

Cognitive Reflection

Rockets and space stations are very expensive. How
much do they cost? How can we justify spending mil-
lions on space when people are starving on our planet?
Provide the pros and cons in a double-spaced, one-
page typed paper (RST.9-10.1).

Handout 6.9A. Space Station Rubric

Name of team: _____

1. A model or drawing was provided.

Poor	Average	Good	Excellent
1 2	3 4 5	6 7	8 9 10

2. Attempts at research was provided.

Poor	Average	Good	Excellent
1 2	3 4 5	6 7	8 9 10

3. Team collaborated well.

Poor	Average	Good	Excellent
1 2	3 4 5	6 7	8 9 10

4. Team could explain how engines work in a clear and
correct manner.

Poor	Average	Good	Excellent
1 2	3 4 5	6 7	8 9 10

5. Team stayed on task.

Poor	Average	Good	Excellent
1 2	3 4 5	6 7	8 9 10

Total: _____

Keep, Retain, and Generalize

How do space engineers work in tandem with others in
specialized fields? Do aerospace engineers, aeronautic
engineers, and scientists do the same thing? Discuss with
a partner how these are the same and how they are different. Make a list of each job description (RST.9–10.1).

LESSON 6.10. BIOFUELS

Back to the Future by George Gipe, based on the screenplay by Robert Zemeckis and Bob Gale (1985)

Summary. Marty McFly, a 17-year-old high school student, gets lost in 1955 by accident, 30 years back in
time. The story takes place in 1985. Marty McFly, a
mild-mannered high school student, was interested in
amplifiers and stopped by Dr. Emmett L. Brown's laboratory to play with the amplifier. He received a message
from Dr. Brown that invited him to help with his new-
est invention, a time machine made out of a DeLorean
sports car that can travel through time instantaneously
when it reaches a speed velocity of 88 mph. His doctor
friend helps him to find his way back to 1985.

Application

Goal. Students will create and explore a vision of future technology by combining their imaginations with
the tools of engineering.

Objectives

- Apply understanding of plant growth to design a
biofuels plant and the biofuels process.

- Discuss the effect an experimental variable (of their choice) has on plant growth.
- Use vocabulary related to biomass, biofuels, and the plant energy process, including photosynthesis and transpiration.
- Plan discussion group for comprehensive questions, self-assessment test, and problem solving.
- Identify agricultural crops and byproducts suitable for renewable energy production and alternative energy sources.
- Understand slang and appropriate time and usage.
- Define science terms associated with *Back to the Future*.
- Provide examples of usage of biofuels for cars.

Standards. To view the Common Core Standards that correspond with this lesson, please visit the *STEAM Meets Story* page on www.tcpress.com and click on the Resources tab.

Teaching Strategy

Introduction. *Back to the Future* centers around a car and finding alternative fuel sources to travel back and forth to the past and future. There are many interventions and innovations that result from students' thinking and problem solving to answer questions. What can happen to a car like the one in the story? Envision the future. What kinds of fuel can we have? The time machine car had a flux capacitor. What kinds of fuels will we need in the future because our oil supply is limited? This lesson is designed around how biofuels are used in transportation. Students will watch the movie *Back to the Future* or read the book. Student will answer conversation questions (Literacy.RST.9–10.2):

1. What are biofuels?
2. Can biofuels be used in cars?
3. Which biofuels are used for transportation?
4. What are examples of biofuels?
5. Why don't we use biofuels?

Essential vocabulary. Define the following vocabulary words (Literacy.RST.9–10.4.):

Biofuels
Biomass
Photosynthesis
Transpirations
Renewable energy

Materials

- Two planting containers (e.g., plastic cups)

Handout 6.10A. Changing Vegetables to Fuel

Record your data and results in the table.

Day	Experiment 1	Experiment 2

- Six seeds (e.g., lima beans, peas, broad beans, etc.)
- Two cups of soil
- A glass bottle or jar
- Options for experimental variables (e.g., two different kinds of soil, fertilizer, sunlight filters heat lamps)
- Vegetable to Fuel worksheet (*Handout 6.10A*)
- Computer access to watch the biofuels video
- *Back to the Future* book/movie (and video; https://www.youtube.com/watch?v=Q2zLXaA1J_4)

Time. Four to 5 class periods

Trial/Try Out

1. Why does the United States need to develop alternative energy sources? Divide students into small groups and give each group 5 minutes to list as many answers as they can to this question. Guide them to understand that there are many important environmental, political, social, and economic reasons the United States needs to develop alternative energy sources (Literacy.RST.9–10.5).
2. Have students share what they already know about biofuel, the science behind it, and its benefits and trade-offs. Based on that list, have students create a list of questions regarding what they would like to learn about biofuels. Have students read Luterbacher and Luterbacher's

(https://kids.frontiersin.org/article/10.3389/frym.2015.00010) article and then answer the following questions. (Literacy.RST.9–10.3, Literacy.RST.9–10.10):

- What materials (biomass) can be used to create biofuel?
- What are the most common types of biomass? What is the science behind the conversion of biofuels?
- Do we have an abundant supply of biomass in the United States? If so, where? What are the economic, environmental, political, and social benefits of using biofuel as an energy source?
- What are the economic, environmental, political, and social trade-offs of using biofuel as an energy source?
- What is the food versus fuel debate related to biofuels, and why is it significant?
- Are biofuels a viable alternative to petroleum-based fuels?
- What are advanced biofuels?

3. Place the samples of biomass materials in front of students (see materials list) and tell them they are going to convert biomass to liquid fuels (Literacy.RST.9–10.1).

- Search online for videos and other resources on converting biomass to liquid fuels and discuss any questions students might have.
- After gathering information on how to make the transformation, students will get in groups of four to design an experiment to compare the amount of ethanol produced in the fermentation of various biomass materials.

Assessment

Write a one-page paper on the need for the United States to have alternative fuel sources using the information from the experiment, film, discussion, and reading.

Cognitive Reflection

If you were an engineer designing alternative energy resources, you would need to know where to plant your crops to get the greatest yield of corn to turn into ethanol. With your group, make a list of things you could grow to produce alternative energy sources most efficiently and the most cost-effective ways to continue the growth of the crops.

Keep, Retain, and Generalize

Students will be asked to reflect on previous biofuel/biomass lessons introduced and the role of biomass in the development of civilization prior to the era of fossil fuel exploitation.

REFERENCES

Alston, R. J., & Hampton, J. L. (2000). Science and engineering as viable career choices for students with disabilities: A survey of parents and teachers. *Rehabilitation Counseling Bulletin, 43,* 158–164. https://doi.org/10.1177/003435520004300306

Berry, R. Q., Thunder, K., & McClain, O. L. (2011). Counter narratives: Examining the mathematics and racial identities of black boys who are successful with school mathematics. *Journal of African American Males in Education, 2*(1), 1–14.

Cameron, J. (Director/Screenwriter), & Hurd, G. A. (Producer/Screenwriter) (1984). *The terminator* [Film]. Hemdale.

Dashner, J. (2013). *The maze runner.* Chicken House.

Dick, P. K. (1956). *Minority report.* Fantastic Universe.

Gipe, G. (1985). Back to the future. Berkley Books.

Jennings, S., McIntyre, J. G., & Butler, S. E. (2015). What young adolescents think about engineering: Immediate and longer lasting impressions of a video intervention. *Journal of Career Development, 42*(1), 3–18. https://doi.org/10.1177/0894845314555124

LaForest, R. K., Gherasoiu, I., White, D., & Efstathaidis, H. (2020, March), *P-TECH: A new model for an integrated engineering technology education.* (Paper presented at St. Lawrence Section Meeting, Ithaca, NY). https://peer.asee.org/33840

Lob, J. (2014). *Snowpiercer.* Titan Comics.

Mantzicopoulos, P., Samarapungavan, A., & Patrick, H. (2009). We learn how to predict and be a scientist: Early science experiences and kindergarten children's social meanings about science, *Cognition and Instruction, 27,* 312–369.

Morales, E. E. (2010). Linking strengths: Identifying and exploring protective factor clusters in academically resilient low-socioeconomic urban students of color. *Roeper Review, 32*(3), 164–175.

National Science and Technology Council. (2018). *Charting a course for success: America's strategy for STEM education.* https://www.whitehouse.gov/wp-content/uploads/2018/12/STEM-Education-Strategic-Plan-2018.pdf

Perry, S. (1994). *Men in black.* Marvel Comics.

Radunzel, J., Mattern, K., & Westrick, P. (2016). The role of academic preparation and interest on STEM success *(ACT Research Report Series).* https://www.act.org/content/dam/act/unsecured/documents/5940-Research-Report-2016-8-Role-of-Academic-Preparation-and-Interest-on-STEM-Success.pdf

Rodenberry, G. (1979). *Star Trek—The motion picture: A novel.* Pocket Books.

Roddenberry, G. (1997). *Star Trek: The Motion Picture.* Fandom.

Romance, N. R., & Vitale, M. R. (1992). A curriculum strategy that expands time for in-depth elementary science instruction by using science-based reading strategies: Effects of a year-long study in grade four. *Journal of Research in Science Teaching, 29*(6), 545–554.

Schwartzbach-Kang, A., & Kang, E. (2018). *Using science to bring literature to life: Combining science and literature can help students engage more deeply with both subjects.* Edutopia. https://www.edutopia.org/article/using-science-bring-literature-life

Shea, D. L., Lubinski, D., & Benbow, C. P. (2001). Importance of assessing spatial ability in intellectually talented young adolescents: A 20-year longitudinal study. *Journal of Educational Psychology, 93*, 604–614.

Shi, Y. (2018). The puzzle of missing female engineers: Academic preparation, ability beliefs, and preferences. *Economics of Education Review, 64*, 129–143.

Spillane, N. K., Lynch, S. J., & Ford, M. R. (2016). Inclusive STEM high schools increase opportunities for underrepresented students. *Phi Delta Kappan, 97*(8), 54–59.

U.S. Department of Education. (2017). *Science, technology, engineering and math: Education for global leadership.* https://www.ed.gov/stem

Wai, J., Lubinski, D., & Benbow, C. P. (2009). Spatial ability for STEM domains: Aligning over 50 years of cumulative psychological knowledge solidifies its importance. *Journal of Educational Psychology, 101*, 817–835. https://doi.org/10.1037/a0016127

Was, C. A., Al-Harthy, I., Stack-Oden, M., & Isaacson, R. M. (2009). Academic identity status and the relationship to achievement goal orientation. *Journal of Research in Educational Psychology, 7*(2), 627–652.

Wei, X., Yu, J. W., Shattuck, P., & Blackorby, J. (2017). High school math and science preparation and postsecondary STEM participation for students with an autism spectrum disorder. *Focus on Autism and Other Developmental Disabilities, 32*(2), 83–92.

Wei, X., Yu, J. W., Shattuck, P., McCracken, M., & Blackorby, J. (2012). Science, technology, engineering, and mathematics (STEM) participation among college students with an autism spectrum disorder. *Journal of Autism and Developmental Disorders, 43*, 1539–1546.

Weir, A. (2011). *The martian.* Crown.

Wu, I. C., Pease, R., & Maker, C. J. (2019). Students' perceptions of a special program for developing exceptional talent in STEM. *Journal of Advanced Academics, 30*(4), 474–499.

CHAPTER 7

Technology and Literature

Jugnu Agrawal, Gary Hoag, and Wen-Hsuan Chang

Literature provides rich opportunities to teach critical technology thinking beyond the typical dystopian novels or Frankensteinian warnings of out-of-control creations. (King, 2018)

The technological use divide continues between students who engage with technology and use it to learn versus students who know how to program computers (U.S. Department of Education, 2017). According to the Office of Education Technology (2017), though there has been an increase in the number of schools who have access to the Internet in classrooms, the gap between high-income and low-income families has been widening. This divide makes it challenging for students from lower income families to explore STEM careers—especially those related to technology. Students need opportunities to engage with technology as more than simply consumers from an early age.

When backed by administrative support, technology can provide equity and accessibility to students from ethnically diverse backgrounds by removing barriers and can reduce the achievement gap. Teachers must demonstrate proficiency and comfort with the use of technology to make it truly accessible for students.

TECHNOLOGY-RELATED CAREERS

With appropriate support both administratively and in the classroom, students have the opportunity to explore multiple careers in the technology field. STEM careers are numerous and varied. The following list of technology careers is far from exclusive; it provides a list of responsibilities as well (Doyle, 2019).

- ***Software developers*** understand and gather information regarding the needs of the end users and then either upgrade an already existing system or design, test, or build something new to meet those needs.
- ***Information security analysts*** develop security standards and practices for companies. They protect the companies' hardware and software from cyberattacks and hackers.
- ***Web developers*** design, create, and maintain websites of the companies to maintain web presence and the public face. They also collaborate with other stakeholders to update the content and information on the website.
- ***Database administrators*** organize, store, and secure everything from financial information to shipping records using software. They create, maintain, update, and secure the databases. They might also merge the old databases with the new ones.
- ***Information security analysts*** protect companies from cyberattacks. They monitor a company's network for security breaches and investigate any breaches or potential breaches that might occur.
- ***Computer programmers*** write codes for computers using languages like Java, C+++, and so forth. They perform several tests on the software, try to locate and anticipate problems, and debug the code to increase accuracy and efficiency.
- ***Computer systems analysts*** act as a liaison between the technical and business sides of a company. They communicate the business requirements with the information technology department and contribute to the overall efficiency of the business. They oversee the project from start to finish.
- ***Computer systems administrators*** keep the computers updated and running smoothly. They install hardware and software programs and maintain the network security and safety

of different companies. They might also train people in different programs.

TEACHER APPLICATION TIPS

This chapter presents a unit with bundled lessons. Why? Because teaching technology without it being immersed within another form of study can be tiresome and even boring. *Hidden Figures* will be used sequentially to enhance skills that the students will learn.

Literature as a tool to introduce technology topics to students helps them overcome their hesitation exploring technological advances. Students need to become critical thinkers and explorers of technology to move beyond just knowing how to use a smartphone or a computer. Teachers can find examples of technological innovations in texts they already use to *intentionally* help their students develop a critical view of learning about digital material. Using literature as a framework will help the students make stronger connections with the technology and observe changes in the way we live.

Culturally relevant literature increases engagement and encourages teachers to use cultural references in the educational environment, which increases students' comfort and confidence level. *Hidden Figures* is an example of culturally relevant text. It is a true story of three African American women who worked at the Langley Memorial Aeronautical Laboratory in Hampton, Virginia. The film version of the story begins around the time the Soviet Union launched Sputnik, the first man-made satellite to orbit the Earth. Jim Crow laws that rigidly demanded strict social separation based on race were entrenched in the government of many states in the United States. Virginia was a segregated state. Langley lab hired African American women mathematicians as human computers; however, they were placed in a separate room to work apart from the White women (Shetterly, 2016).

Hidden Figures, which appeared as a text in 2016 and as a film a year later, covers a number of geographical, geopolitical, STEM vocabulary, and research issues that are intertwined with race relations during the civil rights era. Dorothy Vaughn, Katherine Johnson, and Mary Jackson were leaders who would help send the first American astronaut in space, but they faced huge challenges at Langley. The story highlights the struggles of these three women as the debate between violent and peaceful change was actively and increasingly discussed in the African American community.

SUMMARY OF THE UNIT

The current chapter provides a total of 10 technology-based lesson plans. All the lessons will explore digital tools with the integration of literature and technology within *Hidden Figures*. Scaffolds for making content accessible for English language learners (ELLs) and students with disabilities are included in the lesson plans. The lessons incorporate cooperative learning strategies and presentation skills to assist students with 21st-century skills for the workplace.

LESSON 7.1. IDENTIFYING CREDIBLE SOURCES

Application

Goal. Students will identify and critically evaluate the credibility of online sources.

Objective. Given sources to examine and identify, students will be able to critically evaluate the credibility of the information/sources available online.

Standards. To view the Common Core Standards that correspond with this lesson, please visit the *STEAM Meets Story* page on www.tcpress.com and click on the Resources tab.

Teaching Strategy

Introduction. Students will watch the clips about the movie *Hidden Figures* and read excerpts from the book. Why were these figures hidden? What sets the heroes of *Hidden Figures* apart (https://www.theatlantic.com /entertainment/archive/2017/01/hidden-figures-review /512252/)? What does this have to do with reporting events or news? Teachers can share the video "How False News Spreads" (https://www.youtube.com /watch?v=cSKGa_7XJkg) or a video of their choice to introduce the topic of authentic versus unreliable sources. Then ask students what connections they made with the video. Ask them to share examples. Discuss how access to information has changed and why is it important to be good consumers of information. Just because the information is on the Internet, doesn't mean that it is true.

Materials

- Authentic or Unreliable? worksheet
- Exit ticket
- Book and video clips

Time. 45–60 minutes

Essential Vocabulary

- Digital literacy
- Credibility
- Evaluation
- Consumer
- Reliable
- Authentic

Trial/Try-Out

Teachers will model methods to differentiate between authentic versus unreliable information using criteria such as checking the website or the source, researching the author, reading beyond the headline, and critically evaluating the supporting evidence stated in the article or the source. Students will work with a partner to evaluate the credibility of the articles/news of their choice and complete the assigned worksheet titled "Authentic or Unreliable?" Students will identify two advertisements and two news articles from the website and state reasons to support their classification of the text as authentic or unreliable. Students will generate a list of five to six websites for fact-checking and share it with peers. At the end of the class, they will have generated a list of resources to check credibility (RL.6.7, RL.7.7, RL.8.7, RL.9–10.7, RL 11–12.1).

Assessment

Student responses to the worksheet will be graded as beginning, emerging, or mastery and will be used for formative assessment (*Handouts 7.1A, B, and C*) (RL.6.7, RL.7.7, RL.8.7, RL.9–10.7, RL 11–12.1).

Cognitive Reflection

Students will complete the exit ticket for the following prompt: Think of recent news that you read.

Handout 7.1A. Authentic or Unreliable?

Visit the website for USA Today (usatoday.com) or the website of another news source. Identify four different pieces of information and complete the following.

1. Title of the news article/advertisement

   Is the source authentic or unreliable?

   State the reason to support your claim.

2. Title of the news article/advertisement

   Is the source authentic or unreliable?

   State the reason to support your claim.

3. Title of the news article/advertisement

   Is the source authentic or unreliable?

   State the reason to support your claim.

4. Title of the news article/advertisement

   Is the source authentic or unreliable?

   State the reason to support your claim.

Handout 7.1B. Scoring Rubric

Skill level	Criterion	Notes
Beginning	Student identified some pieces of information but provided little to no evidence to support their claim.	
Emerging	Student identified all pieces of information and provided some evidence to support their claim.	
Mastery	Student identified all pieces of information and provided reasonable evidence to support their claim.	

Describe the news article. Did you think it was authentic? Provide reasons to support your thinking (*Handout 7.1D*) (RL 11–12.1).

Handout 7.1C. Examples for Teacher Demonstration

Share the news article titled "Fake News Onslaught Targets Pizzeria as Nest of Child-Trafficking" (https://www.nytimes.com/2016/11/21/technology/fact-check-this-pizzeria-is-not-a-child-trafficking-site.html). Discuss that the article is authentic, but the information to which the individual in the article responded was unreliable. Discuss how relying on or believing unreliable news can get people in trouble.

Show an article on Wikipedia (https://en.wikipedia.org/wiki/Portal:1960s). Demonstrate how anyone can edit the information listed on Wikipedia. Discuss if the information is authentic or unreliable. Emphasize that the information might be accurate but that students should use other sources to check the accuracy of information.

Keep, Retain, and Generalize

Why is it important to check the authenticity of the information? Discuss with a family member, note their response, and bring it to class (*Handouts 7.1A, B, and* C) (RL.6.7, RL.7.7, RL.8.7, RL.9–10.7, RL 11–12.1).

Handout 7.1D. Evaluate and Summarize

Title:

Source (Name of the website/url):

Describe the news article in your own words:

Did you think it was authentic or unreliable? Provide reasons to support your claim:

LESSON 7.2. THE USE OF CITATIONS AND PRESENTATIONS

Application

Goal. Students will learn to cite different sources to prepare a presentation.

Objectives

- Given specific topics, students will be able to research sources and cite them with 80% accuracy.
- Students will identity plagiarism and cite the different sources and photographs using MLA format with 100% accuracy.
- Students will make a presentation from the research they performed.

Standards. To view the Common Core Standards that correspond with this lesson, please visit the *STEAM Meets Story* page on www.tcpress.com and click on the Resources tab.

Teaching Strategy

Introduction. Show a video clip on plagiarism from Saturday Night Live (https://www.youtube.com/watch?v

=yDxN4c_CmpI). Discuss the following questions: What happened in the video and how did it impact student grades? What were some examples of plagiarism you saw in the video? What might be some strategies to avoid plagiarism (WHST 6–8.8, 9–10.8, 11–12.8)?

Materials

- Various videos
- Computer with speakers
- MLA guide
- MLA formatting cheat sheet (*Handout 7.2B*)
- MLA citations sheet for student practice (*Handout 7.2C*)
- Exit ticket
- Powerpoint or Prezi
- Book

Time. Two 45–60 min class periods

Essential Vocabulary

- Plagiarism
- Citation, Modern Library Association (MLA)

Trial/Try-Out

1. Teachers will show students how to cite information and pictures using MLA format. This will be followed by guided practice where students will review several citations and analyze them for correctness. They will provide the correct citations (*Handouts 7.2A and B*) (WHST 6–8.8, 9–10.8, 11–12.8).
2. For independent practice, students will read the short story "An Occurrence at Owl Creek Bridge" (https://americanliterature.com/author/ambrose -bierce/short-story/an-occurrence-at-owl-creek -bridge) and paraphrase the information about the main character, use direct quotes from the story to explain the setting, and rewrite the ending based on their opinion. Students will cite the information using MLA format. Students will share their endings with a partner (*Handout 7.2B*) (WHST 6–8.8, 9–10.8, 11–12.8)

Technology. Students will choose one of the three women in *Hidden Figures* and prepare a short presentation using MLA formatting to make citations.

Handout 7.2A. Exit Ticket

Three sources of credible information

Two things they learned

One question they might still have

Students can use PowerPoint or Prezi and can include text, art, animation, and audio and video elements (RI 5, RI 7, SMP 3).

Handout 7.2B. MLA Citation Cheat Sheet

Source	Directions for citation	Example
In-Text Citations		
Not citing within the paragraph	"When citing exactly from the book/periodical/website, use quotation marks" (Author's last name page number). When paraphrasing, follow the same directions except no quotation marks.	"As a college graduate and a teacher, she stood near the top of what most Negro women could hope to achieve" (Shetterly 10).
Citing within the paragraph	Provide author's name, title of the book/periodical/website, and page numbers within the paragraph.	Margot Lee Shetterly describes the role of the teachers on page 10 of the book *Hidden Figures*. "Teachers were called upon to do whatever was necessary to keep the schoolhouses clean, safe, and comfortable for pupils."
Reference List Citations		
Book	Author's Last name, First name. *Title of Book.* Publisher, date.	Shetterly, Margot L. *Hidden Figures: The American Dream and the Untold Story of the Black Women Mathematicians Who Helped Win the Space Race.* William Morrow, 2016.
Newspaper/magazine	Author's Last name, First name. "Title of Article." *Title of Magazine/Periodical*, Day month year, page numbers.	Griffin, Chante. "The Benefits of Summer Programs and How to Pay for Them." *Overnight Summer Programs*, 2020, pp. 13–14.
Website	Author's Last name, First name. "Title of the Web Page." Title of the website, Day month year.	Sanchez, Chelsey. "AOC Likens Representative Ted Yoho's Insults to the Toxic Culture That Enables Men to Abuse Women." *Harper's BAZAAR*, July 23, 2020. https://www.harpersbazaar.com/culture/politics/a33404950/aoc-ted-yoho-apology/

Handout 7.2C. MLA Citation Worksheet

Review the following MLA citations and make corrections as needed:

In-Text Citation

Margot Lee Shetterly describes the role of the teachers on page 10 of the book *Hidden Figures*. "Teachers were called upon to do whatever was necessary to keep the schoolhouses clean, safe, and comfortable for pupils."

"When Uncle Mac slowed down from the sharp turn into the Tern Manor private road, Sara clutched the basket handles tighter" (Norton 9).

"As a college graduate and a teacher, she stood near the top of what most Negro women could hope to achieve" (*Shetterly* 10).

Reference List

Griffin, Chante. *"The Benefits of Summer Programs and How to Pay For Them." Overnight Summer Programs.* 2020: 13–14.

Imperato, Cara. "Five Majors That Guarantee Employment After Graduation." College Prep. 2020: 14–15.

Norton, A. "Steel Magic." New York: Open Road Media Teen & Tween, May 22, 2018.

Assessment

Student work samples will be evaluated for MLA citations of content and pictures and will serve as the progress monitoring sample for the class (*Handouts 7.2B and C*) (WHST 6–8.8, 9–10.8, 11–12.8).

Cognitive Reflection

Students will complete an exit ticket with a 3-2-1 structure: three sources of credible information, two things they learned, and one question they might still have (*Handout 7.2A*) (WHST 6–8.8, 9–10.8, 11–12.8).

Keep, Retain, and Generalize

Read an online article and write a citation for the online source using MLA format. Prepare a one- or two-slide PowerPoint (RI 5, RI 7, SMP 3, WHST 6–8.8, 9–10.8, 11–12.8).

LESSON 7.3. MAKING USE OF DIGITAL MEDIA

Application

Goal. Students will integrate information presented in different media or formats (e.g., visually, quantitatively) as well as in words to develop a coherent understanding of a topic or issue and make strategic use of digital media (e.g., textual, graphical, audio, visual, and interactive elements) to enhance understanding of findings, reasoning, and evidence and to add interest.

Objectives. Students will be able to use at least five forms of digital media with 80% accuracy.

Standards. To view the Common Core Standards that correspond with this lesson, please visit the *STEAM Meets Story* page on www.tcpress.com and click on the Resources tab.

Teaching Strategy

Introduction

1. In this lesson, students will research different technologies, digital tools, and inventions. In this unit, in addition to digital tools, students will explore the contributions of some of those characters. Additionally, the teacher will introduce the book and movie *Hidden Figures*. Since America has a long history of discrimination and segregation that many brave people overcame, we will use this platform to discuss these technologies. The book is set in Hampton, Virginia, which was a segregated state before the Civil War. Ask students the following: What do you think the phrase *hidden figures* refers to in the title of the book/movie?

2. Students will then use Padlet to create an online Post-it that can be shared with others. Students can open it on their smartphone or computer. Everyone can see what is being shared in real time and all the ideas gather on the teacher board immediately. Teachers will engage the students in class discussion regarding *Hidden Figures*.

Materials

- Story maps (*Handout 7.3D*)
- Padlet, Quizlet, vocabulary.com, www.etymonline.com, WordsAlive map
- *Hidden Figures* book and movie
- Progress monitoring checklist
- Exit ticket (*Handout 7.3A*)

Time. 45–60 minutes

Essential Vocabulary

Technology terms: Search engine, microblogging, content marketing, website optimization
Literary terms: Setting, characters, exposition, external conflict, rising action, protagonist, and antagonist
Content vocabulary: Proximity, stereotypes, area rule, transonic, supersonic, aerodynamics

Trial/Try-Out

Introduce the key vocabulary for the story elements using *Handout 7.3A* and have students use the digital tool of their choice from the list or paper and pencil to create vocabulary cards that include the following details (RST 6–8.4, L7-8.4c):

- Origin of the word
- Part of speech
- Definition of the morphemes
- Illustration
- Use in a sentence

Handout 7.3A. Morpheme Card

Origin of the word	
Definition of the morphemes	
Part of speech	
Illustration	
Use in a sentence	

Technology. Students will use technologies to present their findings such as PowerPoint or Prezi. Students will define and determine how virtual reality, augmented reality, and artificial intelligence (AI) can be used in a 5-minute presentation to the class on their chosen invention and include visuals (videos, pictures, etc.). They will use what they have learned to provide a summary of *Hidden Figures* and the excerpts of the book or movie that were beneficial to them (WHST 11–12.6).

Handout 7.3B. Frayer Model Template

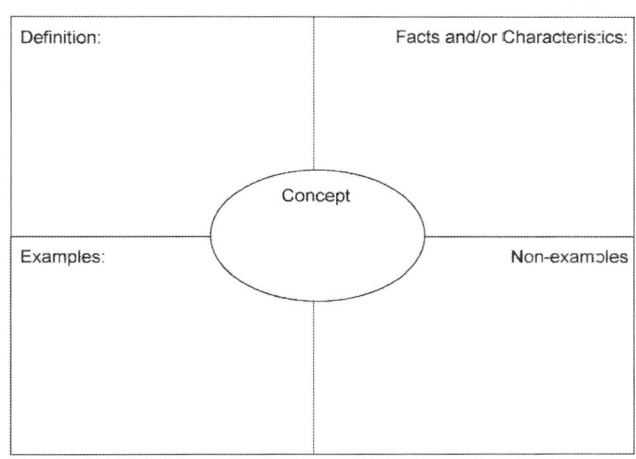

Handout 7.3C. Example of Flip Morpheme Card

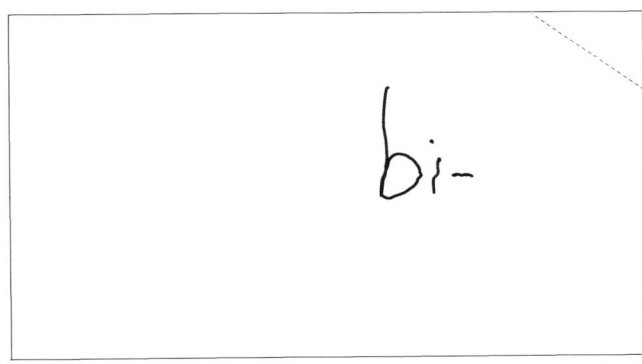

Additional exercises. Teacher will show the film trailer (https://www.youtube.com/watch?v=5wfrDhgUMGI) and use the see, think, and wonder routine. Students will write a journal entry in response to the following questions and include examples from the trailer and text to support their thinking: What do you see? What do you think about that? Does it make you wonder (RI 9–10.7)?

Assessment

Using the rubric in *Handouts 7.3A, B, and C,* students' use and knowledge of digital tools and vocabulary and their journal entry will be assessed. Teachers will check to ensure that all students have chosen a technology or digital tool for their presentation and have used appropriate vocabulary and journal entries.

Cognitive Reflection

Exit ticket: Students reflect on the following prompt: Name someone who is a hidden figure in your life and describe why.

Keep, Retain, and Generalize

Interview someone in your household (parent or grandparent) and ask them who has inspired them and why.

Handout 7.3D. Story Map

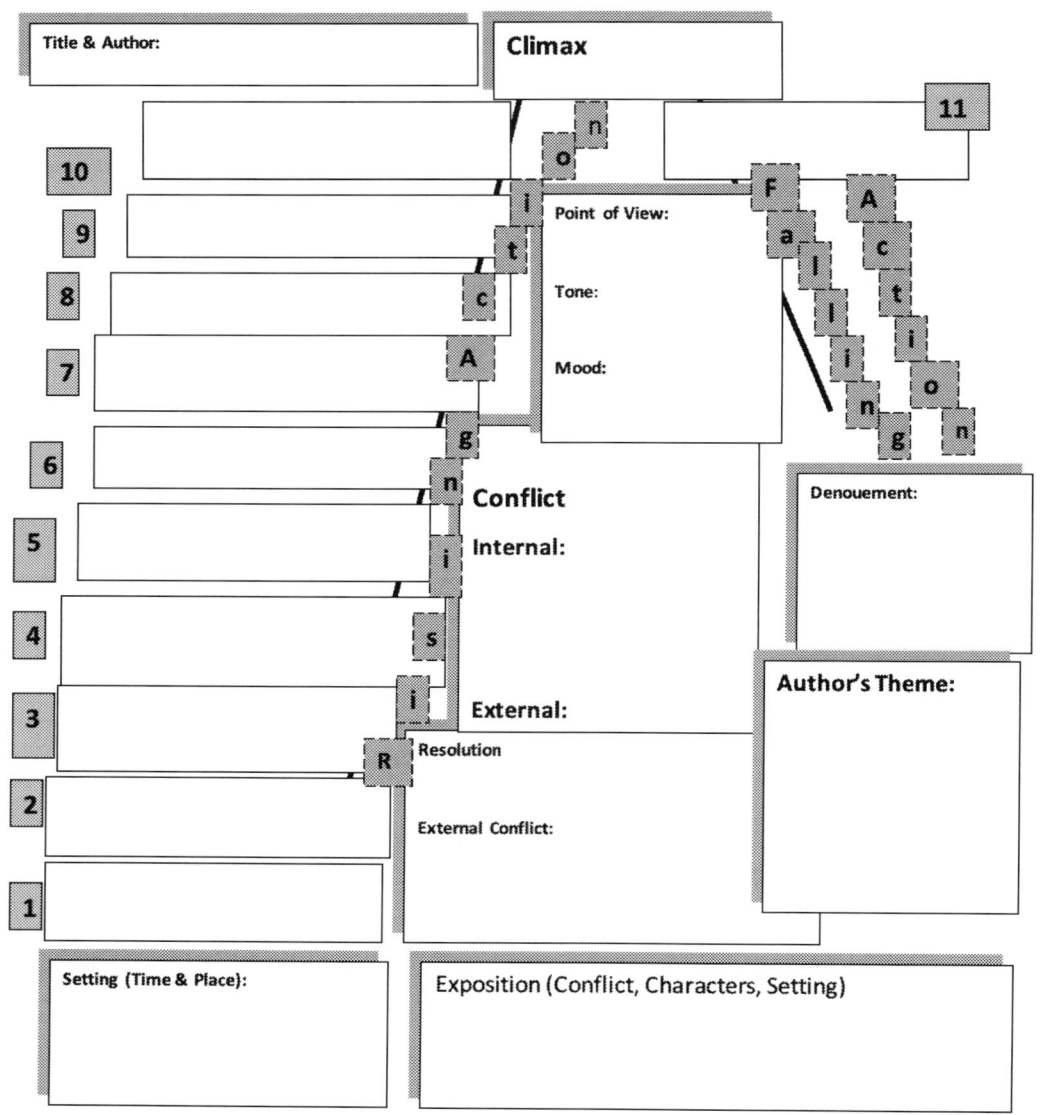

LESSON 7.4. USING TECHNOLOGY TO UPDATE WRITING PRODUCTS

Application

Goal. Students will use technology to publish and be able to analyze text and script.

Objectives

- Students will use technology, including the Internet, to produce, publish, and update individual or shared writing products in response to ongoing feedback, including new arguments or information.
- Students will analyze the extent to which a filmed or live production of a story or drama stays faithful to or departs from the text or script and evaluate the choices made by the director or actors.

Standards. To view the Common Core Standards that correspond with this lesson, please visit the *STEAM Meets Story* page on www.tcpress.com and click on the Resources tab.

Teaching Strategy

Introduction. Students will take a deeper look at the book and learn about and analyze different technologies and inventions.

From the movie *Hidden Figures*, Katherine Johnson gives a speech about how she has to go half a mile to use the bathroom (https://www.youtube.com/watch?v=hNK8FCFpmm4).

- What was life like for African American women in 1957? Discuss with a partner and share with the whole group.

Cooperative learning (timed pair-share): In pairs, students share with a partner for a predetermined time while the partner listens. Then partners switch roles (WHST 11–12.6, RL.8.7).

Materials

- Graphic organizers
- Story maps (students modify versions for differentiation)
- Vocabulary organizers (*Handout 7.4B*)
- Approved list of websites: Padlet, Quizlet
- *Hidden Figures* (leveled text), progress monitoring checklist, and exit ticket

Time. Two class periods, 45–60 minutes each

Essential Vocabulary

Literary terms: Exposition, external, internal conflict
Content vocabulary: Serendipity, code shifting, turbulence, vortex

Trial/Try-Out

1. *Story map:* Based on the reading, students try to complete parts of the story map by writing (or typing) information for characters, exposition, and external and internal conflicts (RL 11.2, 4).
2. *Historical research:* Students must research racial and political situations in 1957 and the years immediately after and reflect on how the beginning of the book introduces these themes. What was the African American community of the protagonists like in the late 1950s and early 1960s? What does the narrator say about the African American community in which she grew up in the film? Complete this writing assignment in Google Classroom (WHST 6–8.7).

Technology. Students will present the technologies they chose. Students will receive a written list of topics to take notes.

Topics on 1950s Technology (*Handout 7.4A*) (WHST 11–12.6)

- How was IBM created and developed?
- How much did World War II and the early Cold War influence technology?
- Who were other scientists and mathematicians (besides those featured in *Hidden Figures*)? What did they contribute?
- After the presentations, students will use their notes to complete the exit ticket.

Assessment

1. *Formative:* Do students understand how to use the story map?
   Summative: Did they complete the story map accurately?

2. *Formative:* What knowledge do students bring to the class about Hampton, Virginia, in 1957?
   Summative: Are students able to say why this location is important to the story?
3. *Formative:* Do students have an understanding of technology in 1957 versus today?
   Summative: Are students able to understand what the computers did for work at that time?

Cognitive Reflection

Exit ticket. Students will complete a discussion board post with their journal entry:

> Which technology resonated with you most and why? Students will comment on the posts of two other classmates.

Handout 7.4A. 1950s Technology Topics

Lesson Plan 04: 1950s Technology Topics

Do you think we had no technology in the 1950s? Think again!

How was IBM created and devaloped?

How much and in what ways did World War II and the early Col War influence technology?

Who were other scientists and mathematicians of the 1950s (besides those featured in *Hidden Figures*)?	**What did they contribute?**

Handout 7.4B. Flip Morpheme Card

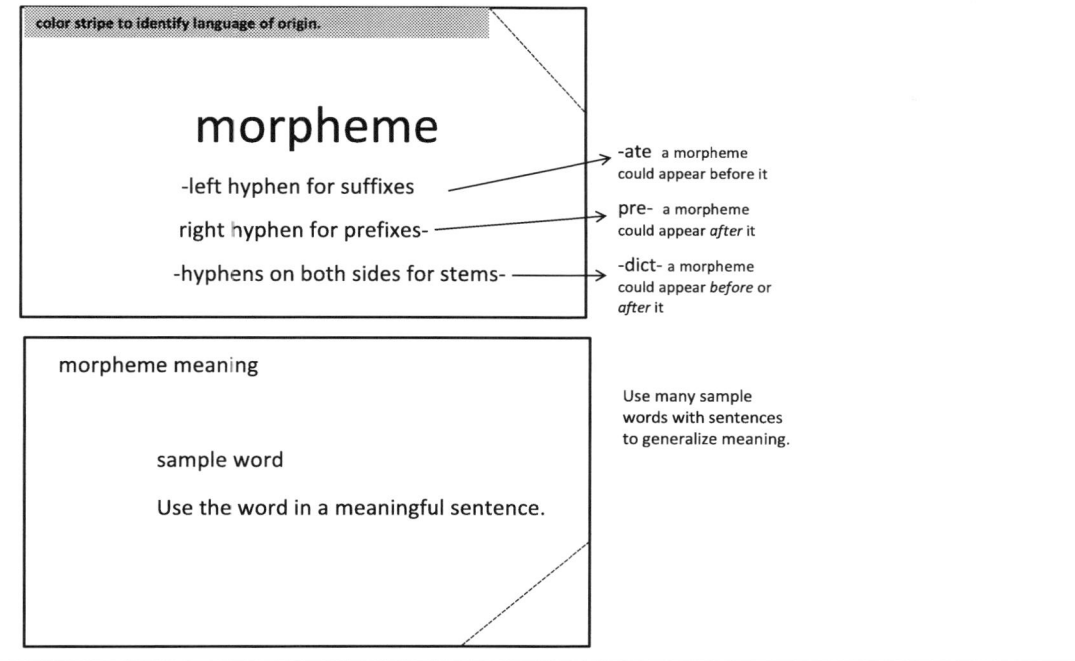

Keep, Retain, and Generalize

Out-of-class assignment. Why is 1957 important? What happened, and how did it change the way we looked at ourselves as a country? Students write their answers. (Possible answers: We changed our education system and provided more funding for education.)

LESSON 7.5. DIGITAL CITIZENSHIP AND TECHNOLOGY INTEGRATION

Application

Goal. Students will discuss tough topics to build positive online communities and stop online cruelty. Students will also analyze text for differences in versions.

Objectives

- Describe digital citizenship and online hate speech and how to combat it.
- Recognize differences among various versions (print versus film) of the book.
- Cite strong and thorough textual evidence to support analysis of what the text says explicitly as well as inferences drawn from the text, including determining where the text leaves matters uncertain.

Standards. To view the Common Core Standards that correspond with this lesson, please visit the *STEAM Meets Story* page on www.tcpress.com and click on the Resources tab.

Teaching Strategy

Introduction. We begin with whole class instruction. Students will see two film clips from *Hidden Figures* (Dorothy taking the library book; https://www.youtube.com/watch?v=ID1iFaWgcIE) and (operating the "computer" later; https://www.youtube.com watch?v=x5GV7ODht78). Students will work in small groups and respond to the following prompts for discussion: What happens to African American women who have mathematical skills? Dorothy takes a book from the "Whites only" library; how does she use it? How do current innovations compare with the innovations in the movie/book?

Students will work in triads and will be assigned the following roles: timekeeper, note-taker, and spokesperson. The timekeeper will keep track of time; note-taker will capture the notes from small group discussion; and spokesperson will share key points from the discussion with the whole class (RL 6.7, 7.7, 9-10.7, 11-12.7, 11-12.1). (*Handout 7.5A*)

Materials

- Movie clips
- Note cards
- Timer
- Story maps
- Peer feedback checklist
- WordsAlive sample

Time. One to two class periods

Essential Vocabulary

- Counter speech
- Extremism
- Hate speech
- Xenophobia
- Protagonist
- Antagonist
- Fortran
- Fluid dynamics
- Nimble
- Semi-autonomous

Trial/Try-Out

1. *Story map:* Review story map literary terms as applied to this book. Ask students the following: Who is/are the protagonist(s)? The antagonist(s)? Does the setting of the story make it more or less remarkable? Students think about their responses to the questions, discuss answers in pairs, and share their own or their partner's answer with the class (RL 6.7, 7.7, 9–10.7, 11–12.7, 11.1).
2. *Research:* Students research historical background and answer the following questions: Does historical and geographical/geopolitical background make a difference to the story? They can use research conducted by others in the group to create a graphical outline or poster describing this issue. Students can use several online sources (RL 6.7, 7.7, 9–10.7, 11–12.7, 11.2).
3. *Technology/writing:* Respond to the following quote from the text/film: "It is not because we wear skirts but we wear glasses." What is the significance of this quote? Please provide feedback on writing using the teacher provided checklist (RL 6.7, 7.7, 9–10.7, 11–12.7, 11.5).

Technology. Digital citizenship: Watch the video on good digital citizenship (https://www.commonsense .org/education/videos/what-is-digital-citizenship). Hate

Handout 7.5A. Writing Checklist

Composition:

- ☐ Topic sentence
- ☐ Supporting details
- ☐ Elaborations (referenced in the text)

Written expression:

- ☐ Sentence variety (simple, complex, compound; sentences begin with different structures such as prepositional phrases and dependent clauses)
- ☐ Active (rather than passive) sentences

Usage/mechanics:

- ☐ Correct spelling
- ☐ Verb tenses
- ☐ Subject-verb agreement
- ☐ Pronoun-antecedent agreement
- ☐ Punctuation
- ☐ Capitalization

speech has existed for a long time, but lately there has been an increase online. Sixty-four percent of teens have encountered it (https://www.commonsensemedia .org/research/social-media-social-life-2018). One form of hate speech is called xenophobia, prejudice against those from other countries. Satirical videos, photos, music lyrics, and blogs have also been used to platform hate speech, and various stereotypes have been used against various religions and ethnic groups. We witness similar treatment in the *Hidden Figures* book. In a group of four, create a video, blog, presentation, or other technology to combat this behavior that will promote positive and strong digital citizenship (*Handout 7.5B*).

Assessment

1. *Formative:* Can students identify main characters and the role played by each?
   Summative: Do students understand the importance of the setting?
2. *Formative:* Can students identify and discuss the setting? How is this setting different from, say, New York?
   Summative: Is students' graphical outline accurate?
3. *Formative:* Do students' writing reflect to whom the character is speaking and why that is important?
   Summative: Do students understand the interplay between sexism and racism?

See *Handout 7.5B* for the rubric for Digital Citizenship.

Handout 7.5B. Digital Citizenship

Name of team: _____

1. Does the piece provide a platform for digital citizenship?

Poor	Average	Good	Excellent
1 2	3 4 5	6 7	8 9 10

2. Does the piece have strong technology quality?

Poor	Average	Good	Excellent
1 2	3 4 5	6 7	8 9 10

3. Did the team collaborate well?

Poor	Average	Good	Excellent
1 2	3 4 5	6 7	8 9 10

4. Is digital citizenship explained in a clear and correct manner?

Poor	Average	Good	Excellent
1 2	3 4 5	6 7	8 9 10

5. Did the team stay on task?

Poor	Average	Good	Excellent
1 2	3 4 5	6 7	8 9 10

Total: _____/50

Cognitive Reflection

Exit ticket. What connections do you make between digital citizenship and your personal life? Write two to three sentences to describe it.

Keep, Retain, and Generalize

Out-of-class assignment: Research the meaning and development of FORTRAN and other computer languages. Report to your group.

LESSON 7.6. WEBCAMS AND IDENTIFYING CENTRAL THEMES

Application

Goals. Students will identify themes and central ideas. Students will also understand the use of webcams.

Objectives

- Students will determine two or more themes or central ideas of a text and analyze their development and how they interact and build on one another.
- Students will compare the text with the movie.
- Students will learn about the experiences provided by the webcam and make a webcam video.

Standards. To view the Common Core Standards that correspond with this lesson, please visit the *STEAM Meets Story* page on www.tcpress.com and click on the Resources tab.

Teaching Strategy

Introduction. Teachers choose the scene in the book and film in which the problem of how to bring the astronaut back from the elliptical orbit is addressed (https://www.youtube.com/watch?v=v-pbGAts_Fg) and also read the pages of the book (pp. 219–222) that include the preparation for the meeting. Students break into triads and complete a Venn diagram that focuses on the differences between the book and film: What is the difference between an elliptical orbit and a parabolic one? Why is this important? Why was this discussion significant to the success of the mission (RL 6.7, 7.7, 9–10.7, 11–12.2)?

Materials

- Morpheme cards/index cards
- Computer with Internet access
- Projector with access to movie clip
- Speakers
- Venn diagram

Essential vocabulary. See *Handout 7.6B.*

Literary terms:
- Climax
- Falling action
- Denouement

Content vocabulary:
- Rotation
- Elliptical
- Gravity
- Heliocentric

- Oblations
- Parabolic
- Protocol

Time. Three class periods

Trial/Try-Out

Introduction

1. Research and project: Students will work in pairs and choose and research a technology topic and relate it to the book for final presentation. Students will use technology tools to prepare their final presentation and produce a model, PowerPoint, outline, poster, or use some other template/media tool. Students will choose from a list of potential topics on technology (*Handout 7.6A*) (RL 6.7, 7.7, 9–10.7, 11.33).
2. Writing/character development: Respond to one of the following using either the book or the film:
   - Which protagonist is the main character? How do we know?
   - Choose a protagonist whose story is impressive to you. What did you find impressive about the story and what did they do in the face of adversity?

Handout 7.6A. Potential Topics for Project

- The way the actual technologies have evolved since the book (e.g., computer science languages, computer sizes, spaceship durability testing processes, etc.)
- How the accessibility of engineering education has changed
- STEM issues related to current or future space projects
- If women and people of color have progressed in STEM occupation and why or why not
- Contributions of minority groups (women, African Americans, etc.) at NASA
- Space station design. Think about the essential needs of the astronauts and issues that may arise. Provide examples of technologies that will help the astronauts.
- Movie with the characters of *Hidden Figures* in today's day and time
- Website using WordPress, Wix, or any other web designing tool. Document America's race to the moon.
- Blog with multiple entries from the perspective of a character from the book/movie (identify the key moments in detail).

- Pick a scene or event from the book/story that was powerful for you and explain why (RL 6.7, 7.7, 9–10.7, 11.1).
3. Content-specific vocabulary: Students use tools for key vocabulary and concepts to create cards for content-specific words: continuum, elliptical, gravity, heliocentric, quantum, parabolic, protocol.
   - Continue building a file on Quizlet or in a vocabulary box of morphemes (and words) related to the story.
   - Create a Quizlet for the key vocabulary and concepts (RL 6.7, 7.7, 9–10.7, 11.3).

Technology. In this lesson, you will learn to use a webcam to take picture and create a video clip (PA Virtual Charter School, 2017) and how to hook your webcam to your phone (https://www.youtube.com/watch?v=3-_pIos5n8s). Because *Hidden Figures* is the focus of this unit, we will take a virtual real-time cyber trip to visit

Handout 7.6B. Essential Vocabulary: Literary and Content Terms Word Bank

Use the context clues to fill in the blanks with the correct vocabulary (some words are not used).

Climax	Falling action
Denouement	Rotation
Elliptical	Gravity
Heliocentric	Oblateness
Parabolic	Protocols

The book *Hidden Figures* concerns the real-life efforts of three African American women who, as human "computers," were responsible in part for the success of the American space program. While dealing with the _____ of segregation—rules that separated and discriminated against them—they nonetheless became successful mathematicians and scientists. An example of these achievements came during the debate on the _____ orbit astronaut John Glenn needed to achieve to re-enter the Earth's atmosphere. Glenn needed to shift his orbit from the oval-shaped _____ orbit his Mercury craft had assumed to avoid skipping into space and joining the Earth in its _____ travels around the sun. Katherine Johnson had to account for numerous variables in her mathematical equations, including the Earth's _____ as it turns and it's _____ (the fact that the Earth is not complexly round). Katherine had insisted on pursuing her personal dream, and the _____ of her story came with the successful return of Mercury 7 to Earth.

the National Aeronautics and Space Administration (https://www.nasa.gov/multimedia/nasatv/iss_ustream.html) at NASA Live (https://www.nasa.gov/nasalive) to see the results of the work done by the women in the movie and the book and visit the International Space Station (ISS). Afterward, in a group of four, you will make a webcam video about the women in *Hidden Figures*. Combine what you find in the movie, the book, and your Internet research to create a webcam video (*Handout 7.6C*) (Communication and Collaboration W 6, W 10 SL 2, SL 5).

Assessment

1. *Formative:* Are students familiar with some of the content specific words?
   Summative: Did they understand the meaning of morpheme and the use of the morpheme cards?
2. *Formative:* Have students developed an understanding of STEM as a multidisciplinary field?
   Summative: Do students understand how race, gender, ethnicity, ELL, and special education issues influence STEM careers?
3. *Formative:* Can students identify main characters and their roles?
   Summative: Can students identify how each character played a role in the success of NASA?

Cognitive Reflection

Exit ticket. Students answer exit ticket questions, focusing on vocabulary in context with seven lesson plan questions.

Handout 7.6C. Webcam Video

Name of team: _____

1. Does the video provide good research?

Poor	Average	Good	Excellent
1 2	3 4 5	6 7	8 9 10

2. Is the technology used correctly?

Poor	Average	Good	Excellent
1 2	3 4 5	6 7	8 9 10

3. Did the team collaborate well?

Poor	Average	Good	Excellent
1 2	3 4 5	6 7	8 9 10

4. Did the team explain *Hidden Figures* women in a clear and correct manner?

Poor	Average	Good	Excellent
1 2	3 4 5	6 7	8 9 10

5. Did the team stay on task?

Poor	Average	Good	Excellent
1 2	3 4 5	6 7	8 9 10

Total: _____/50

Keep, Retain, and Generalize

Out-of-class assignment. Find out when the International Space Station (ISS) will be flying overhead (https://spotthestation.nasa.gov).

LESSON 7.7. DIGITAL FOOTPRINT

Application

Goal. Students will learn about their digital footprint and how that matters to their reputation and making sound judgments on what they place on the Internet and social media.

Objectives

- Given five opportunities, students will be able to name at least five warnings related to their digital footprint with 80% accuracy.
- Given five opportunities, students will make reasoned judgments when deciding on research

and be able to name at least two ways to spot speculative text.

Standards. To view the Common Core Standards that correspond with this lesson, please visit the *STEAM Meets Story* page on www.tcpress.com and click on the Resources tab.

Teaching Strategy

Introduction. We begin with whole class instruction. Play a movie clip about reading redacted information (https://www.youtube.com/watch?v=JAEnv1PvBvw). Also potentially play the scene where they question

Katherine in the office and ask her if she's a spy (https://www.youtube.com/watch?v=eEJWtAMnAlA). Who had access to classified information in the engineering reports? How do we know? Which professions today need security clearance? How do we get it? First, do "Turn Toss": Students toss a ball (paper wad) while answering the question about who had access to classified information. Then, students shift to write their thoughts (RST.6.7, RST.7.7).

Materials

- Story maps
- Computers

Essential Vocabulary

Literary terms

- Climax
- Resolution
- Falling action
- Denouement
- Theme

Content vocabulary

- Contentious
- Circumvent
- Fuselage
- Incisive
- Retrofit
- Synecdoche (see vocabulary.com and Quizlet for more vocabulary; reference etymonline.com to discover morphemes and etymology)
- Digital footprint

Time. Three class periods (45-60 mins each)

Trial/Try-Out

1. Students complete their story maps and present them in their small group with a focus on climax, falling action, tone, mood, point of view (narrator), denouement, and theme (RL 8.9, 9–10.9).
2. Students will work in their assigned pairs for the project and research the content vocabulary words, find other uses of these words, and try to incorporate them in their project. The remainder of the time will be used for work on their long-term research project (RL 8.7, 9–10.7, RL.6.7, RL.7.7).
3. Respond to one of the following:
   a. Which character in the book do you relate to most and why?
   b. If you had to follow in the footsteps of one of the characters in the book's career path, which one would you choose and why? Examples include engineer, mathematician, manager, and military official (RL 11.6).

Technology

1. The women in *Hidden Figures* were very intelligent. They also had a good reputation and became the subject of a book and a movie because of the positive things they did. There are a lot of things we do digitally, and some of it can be problematic. Students will learn about the digital contract and its negative and positive impact. We have 24/7 access to online tools through social media, videos, and various ways to publish. There are benefits and risks involved. You leave a digital footprint when you are browsing and using social media. Your digital footprint is anything you do online, including activities such as posting images, texting, and engaging in other forms of online communication. It is there forever because many websites collect your IP address and you cannot delete what is already posted. Your digital footprint is used to collect personal information about you, for example demographics, religion, political affiliations, or interests. Information can be gathered using cookies, which are small files websites store on your computer after your first visit to track user activity. When you are shopping online, preferences are stored and advertisers can target what kind of ads to send to your browser. They are called cookies. If you are attempting to find a job, future employers can also retrieve information, and they might see something unfavorable that could jeopardize your job search. Read the article "Your Digital Footprint: What Is It and How Can You Manage It?" (https://www.rasmussen.edu/student-experience/college-life/what-is-digital-footprint/). What are some things you can do to manage your footprint? The article lists six things you can do; summarize them.
2. There are some positive things that can come from the digital footprint. A positive online presence, or digital footprint, can boost your career just as a weak one can damage it. Now that you know what not to do, what can you do to boost your career? What things can you do? Hint: Remember your hobbies, things you do for other people and the community, family photos, and other things (*Handout 7.7A*) (RST.6.8.1, RST9-10, RST.6–8.3, RST.9–10.3, RST. 6–8.9, 9–10.9).

Handout 7.7A. Exit Ticket: Matching!

Enter each word's number next to its definition.

1. Contentious   ___ untying the knot

2. Climax   ___ a spindle, a tube

3. Denouement   ___ winding up the plot

4. Circumvent   ___ surround, usually in a hostile manner

5. Fuselage   ___ quarrelsome

6. Retrofit   ___ resolves conflicts

Assessment

1. *Formative:* Do students understand that more than a single event could count as the climax?
   Summative: Have students mastered the vocabulary?
2. *Formative:* Do students recognize the content vocabulary? Are they able to make independent progress on their projects?
   Summative: Are students able to use content vocabulary in sentences?
3. *Formative:* Are students able to write complex sentences?
   Summative: Are they able to make independent progress on the questions?

Cognitive Reflection

Exit ticket. Students answer exit ticket questions focused on matching vocabulary (*Handout 7.7A*).

Keep, Retain, and Generalize

Out-of-class assignment. Explore other ways to make your digital footprint more positive and discuss with a family member. (*Handout 7.7B*)

Handout 7.7B. Digital Footprint

List what you need to do improve your digital footprint. What things do you need to avoid that would hurt your digital footprint?

Things to avoid:

1.

2.

3.

4.

5.

6.

Things to improve:

1.

2.

3.

4.

5.

6.

LESSON 7.8. ASSISTIVE TECHNOLOGY INNOVATIONS

Application

Goal. Students will analyze the author's structure as well as technological issues and learn about some assistive technology devices.

Objectives

- Given a task, students will analyze the technological issues with 90% accuracy.
- Students will name and explain at least two assistive technologies that help persons with disabilities.

Standards. To view the Common Core Standards that correspond with this lesson, please visit the *STEAM Meets Story* page on www.tcpress.com and click on the Resources tab.

Teaching Strategy

Introduction. Students share their responses from the exit ticket for Lesson Plan 7.7 to the question "What was the denouement, and what does it portent for the future?" Cooperative learning: What major American companies developed machines that could replace the work of the human computers? What attitude did that company's characters bring to Langley, and what did our protagonists do about it? What does this teach us about becoming critical consumers of technology? Students are provided the prompts on chart paper, and they are posted in three different areas of the classroom. Students discuss and respond to each prompt in a small group. When the timer goes off, they rotate to the next prompt. Once all groups have responded to all prompts, the last group shares the key points from the chart (RST.6–8.5, RST. 8.6.5, RST. 9–10.5, RST. 11–12.5).

Materials

- Graphical outlines
- Padlet
- Google Classroom
- YouTube videos

Time. Two class periods, 45–60 minutes each

Essential Vocabulary

- Prototypes
- Miniaturization
- Trajectory
- Moore's law
- Patents
- Nominal
- Byte
- Peripheral

Trial/Try-Out

1. *Content vocabulary:* Prototypes, miniaturization, trajectory, Moore's law, patents, nominal, byte, peripheral (https://public.oed.com/blog/words-from-the-1960s/). Students create vocabulary tools to support their learning (RI 8.4, RL.6.7, RL.7.7, RL.9–10.7).
2. How did private corporations assist the space program? Were such companies beneficial for the civil rights movement? What about current technology companies such as Amazon, Google, SpaceX, and so forth? Do they advance civil rights? Provide specific examples to support your conclusions. Students complete the statement, "I used to think . . . , but now I think . . ." (RI 11.7, RL.6.7, RL.7.7, RL.9–10.7).
3. *Students respond to one of the writing prompts:* What issues are presented related to technological unemployment? What is meant by the phrase? Are people of color disadvantaged by technological advancements? How did the African American women avoid losing their jobs? Students can write about the role of private technology companies (W 8.2. a-f, W9-10.2, W11–12.2).

Technology. There are many technological devices we use, such as the Chromebook and iPad. You have been introduced to a lot of technology in this unit, but there is some technology designed to help those with special needs:

- eSight electronic glasses allow the legally blind to "see" without the need for surgery. The glasses contain a high-speed, high-definition camera that captures what is seen and then algorithms enhance the video feed.
- OneNote is a digital note-taking application that has been enhanced for the vision and mobility impaired with simplified navigation controls and consistency.
- Livescribe's Smartpen works like a normal pen but also syncs everything you write to what is being said aloud. If you touch the pen anywhere on your notes, it plays information back. Research Bookshare and ActiveWords and describe their function. Find two other technology tools that will help K–12 students (*Handout 7.8A*).

Assessment

1. *Formative:* Have students mastered the use of various tools to ascertain correct vocabulary meanings and use? *Summative:* Can students use newly acquired vocabulary in sentences?

Handout 7.8A. Assistive Technology Devices

Research and find two technologies that can be assistive.

1.

2.

2. *Formative:* Are students able to research and assess results?
   Summative: Are students able to cite meaningful examples?
3. *Formative:* Are students able to write complete complex sentences and assess issues presented?
   Summative: Are students able to compose a paragraph that meets a teacher-created writing rubric?
   Students will list four technological devices for a score of 100, three for a score of 80, and two for a score of 70.

Cognitive Reflection

Exit ticket. Students select any two vocabulary words from the following list and illustrate them: continuum, elliptical, gravity, heliocentric, quantum, parabolic, protocol, prototypes, trajectory, patents, nominal, byte, peripheral

Keep, Retain, and Generalize

Out-of-class assignment: Follow the NASA Twitter feed and find one that interests you. Come prepared to share with the class.

LESSON 7.9. DIGITAL STORYTELLING

Application

Goal. Students will make a film relating to the book *Hidden Figures* while acquiring an understanding of segregation.

Objectives

- Students will be able to answer two questions related to segregation with 100% accuracy.
- Students will make a film telling the story of *Hidden Figures* from a section of the book not in the movie.

Standards. To view the Common Core Standards that correspond with this lesson, please visit the *STEAM Meets Story* page on www.tcpress.com and click on the Resources tab.

Teaching Strategy

Introduction. Students will watch the court scene with Mary Jackson (https://www.youtube.com/watch?v=btm0uybciPA) and engage in discussion about the following questions: What was the role of historically Black colleges? Did our protagonists have opportunities that they would not have received otherwise? What did they do to receive these opportunities?

Cooperative strategies. Class divides into two-person teams. Each team writes a statement to answer the question; teams present their answers briefly to the class.

- What was the role of historically Black colleges? Did our protagonists have opportunities that they

would not have received otherwise? What did they do to receive these opportunities?
- How was the existence of sororities for Black female students important?
- What role did segregation play in providing or negating opportunities African Americans?

Students will use multiple sources of information to analyze these questions, determine the meanings of the words used in the text, and analyze themes and choices the author made to tell the story (RL.9–7, 10.7, RL. 11–12.7).

Time. Three class periods, 45–60 minutes each

Materials

- Graphical outlines
- Padlet
- Google Classroom
- YouTube videos

Essential Vocabulary

- Aerospace
- Aeronautical
- Cold War
- Rosenbergs
- Supersonic
- Wind tunnel

Trial/Try-Out

1. What do we know about each protagonist and her background? Compare her background in the film with the women's actual biographies (https://

www.nasa.gov/modernfigures/media-resources). Students respond to the discussion boards to answer the question "How is the film version different from reality?" They also respond to two other students' posts (RI.6–7).

2. Watch the videos of some modern figures (https://www.nasa.gov/modernfigures/videos). Imagine an ideal world and write a passage in which those issues don't exist. Be as specific as possible and explain what one would feel/experience in a day at work in the ideal world. Share with the group (RI.6–7, RI.8.7).

3. Students will write about the history of segregation in American history. Additionally, they will write about the fight for equality in one of the following areas: rights of women, equal pay, fight against racism, maternity and paternity leave, LGBTQ, or other (RL.11–12.7).

Technology. Digital storytelling can be used to make films about anything. Students can find a scene in the book *Hidden Figures* that is not in the movie and make a story. Creating stories provides a sense of ownership. Students can demonstrate their creativity while writing the story. Digital story helps with communication because students can help better organize their ideas and narratives and articulate those using digital tools. They can publish it on YouTube. Students work in teams to create their stories, which helps with social collaboration skills (https://www.educationworld.com/how-filmmaking-schools-can-foster-creativity-all-learners) (RL.7.7. RL.9–10.7). Make sure that you include the following for your movie:

- The main point of view and perspective as the driving force
- A key question to garner the viewer's attention and be resolved in the end
- Serious issues that connect well with audiences and can become powerful elements
- Music added as the soundtrack to embellish the story

- Points of the story without unnecessary information
- Rhythm

Assessment

1. *Formative:* Are students familiar with the discussion board concept?
   Summative: Instructors review discussion boards or Padlet posts to determine whether students answered questions or posed queries relevant to the assignment.

2. *Formative:* Do students have a concept of what is meant by an ideal world?
   Summative: Did students select an appropriately related topic for investigation and eventual end-of-book presentation?

3. *Formative:* Do students understand the history of segregation?
   Summative: Were student writing samples appropriately detailed and structured (topic sentences, complex sentence structures)?

4. Were the students able to identify information from the book that was not included in the movie?

Cognitive Reflection

Exit ticket. Who has influenced your life in a way that has allowed you to be successful today? This could be someone you know personally or a historical figure. Share using a bulleted list.

Keep, Retain, and Generalize

Out-of-class assignment: Students listen to "Katherine Johnson: The Girl Who Loved to Count" (https://www.nasa.gov/feature/katherine-johnson-the-girl-who-loved-to-count) and answer this question: Why was her interest in counting important? Ask students to go beyond the actual question and ask themselves how such an interest could impact not only a single life but the life of a project and a nation.

LESSON 7.10. 3D PRINT

Application

Goal. Students will be able to establish rules for collegial discussion, present their project, and provide peer feedback on other student projects. Students are introduced to the fundamental concept and logic necessary to use 3D print.

Objectives

- Students will be able to make 3D prints.
- Students will use 3D prints and other technologies for the last lesson in the unit.

Standards. To view the Common Core Standards that correspond with this lesson, please visit the *STEAM Meets Story* page on www.tcpress.com and click on the Resources tab.

Teaching Strategy

Introduction. Students will watch the movie clip of the girl who loved to count (https://www.nasa.gov/feature /katherine-johnson-the-girl-who-loved-to-count). Students will engage in whole class discussion using the prompt "Define segregation. What role did it play in providing opportunities for African Americans?" For discussion, students will use the stand up, hand up, and pair up structure. For this activity, students will stand up, put their hand up, and find a partner to share and discuss (SL 10.1.b).

Materials

- ATTACK rubric (copies for all student projects), technology for project presentations
- 3D printers
- Computer or laptop
- White sheet or a large sheet of paper to draw the design
- Measuring instruments
- 3D print objects

Time. One class period

Essential Vocabulary

Pair up with a classmate to discuss the meaning of the following words. Also, discuss a situation where you have read, used, or seen these vocabularies (Science and Technical Subjects, RST. 6–8.4).

- Blueprint
- Innovation
- Scan
- Modification
- 3D

Trial/Try-Out

Students will present their plan for final projects for the unit. As students view the presentations, they will complete the rubric related to different aspects of presentation and jot notes to respond to the following questions: How does each presentation relate to the book? What new idea/concept/information did the group present? How does the presentation relate to our time?

Technology

1. What is 3D printing (RST.6–8.4, RST, 9–10.4)?
2. What can a 3D printer make (RST.6–8.7)?
3. What are the key materials for 3D print (RST.6–8.4, RST, 9–10.4)?
4. 3D print is used widely today in school. How could 3D printing be used in your daily life (RST.6–8.7)?
   Online intro websites to 3D printing include MatterHackers (https://www.youtube.com /watch?v=nbIoKadLIuw; https://www.youtube .com/watch?v=8WhqM5dS6uE), Yeggi, and TinkerCad (Science and Technical Subjects, RST.6–8.7).

Additional Exercises

1. What are the differences between 2D and 3D? What else in the classroom could benefit from using a 3D printer (RST.6–8.3, RST.9–10.3)?
2. What is an implication of using a 3D printer from reading the story (RL 6.7, RL7.10)?
3. Design a 3D object related to the book as a group (RST.6–8.3, RST.9–10.3, RST.6–8.7), which you can use to enhance your final presentation.

Assessment

1. Teachers use a rubric to assess student presentations. Students should write notes on their rubric and turn those into the teacher (*Handout 7.10A*).
2. From designing to completing a 3D print object, what types of problems did you encounter? How did you work through the appropriate measures and steps for completing a 3D print project (RST. 6–8.1, RST. 9–10.1; RST 6–8, 3, RST. 9–10.3)?

Cognitive Reflection

1. Whole class discussion after the presentations on the three questions in "Additional Exercises."
2. What other creative ways can you think of to use 3D print (RST. 6–8.1, RST.9–10.1)?

Keep, Retain, and Generalize

1. Has this unit given you new ideas about your dream profession and why?
2. How will knowledge about 3D print help you in another class (RST. 6–8.9, RST.9–10.9)?

Handout 7.10A. Summative Assessment

Circle the category that best reflects the performance of the team.

	Emerging (1 point)	Developing (2 points)	Proficient (3 points)	Exemplary (4 points)
Content	Content lacks a clear point of view and logical sequence of information. Information is inaccurate or missing key points.	Content has a main idea not clearly stated. Some of the information included doesn't seem to fit.	Content is written with a coherent progression of information and the information is accurate.	Content is written clearly and concisely with a sequential order of facts and supporting details.
Presentation Skills	Information was not articulated clearly and did not match the presentation.	Information was articulated but wasn't clear and did not match the presentation.	Information was articulated with a few parts that did not follow the presentation.	Information was clearly articulated and followed the presentation.
Technology	Students used only a basic slideshow with no other technology tools (graphics, media, etc.).	Students used a technology tool and graphics but did not include other media sources.	Students included multiple technology tools including graphics and audio or video.	Students included various technology tools, including graphics, audio, and other media sources, to enhance the information.
Connection with the book	Student presentation did not relate to the book and did not achieve project goals.	Students illustrated some knowledge of the book and considered design elements.	Students' project shows a clear connection with the book and draws parallels at multiple levels, which are clearly explained.	Students bring a variety of considerations and creative elements to a well-organized presentation that relates to the book.
Writing	Presentation needs extensive editing. Multiple errors in grammar, punctuation, spelling; difficult to understand.	Structure is missing. Many errors in grammar, punctuation, and spelling. Presentation needs editing.	Clear, concise, and well written with minor errors in grammar and usage.	Essay structure is clear with appropriate thesis, topic sentences, details, and elaboration. Writing is grammatically appropriate.
3D Object	Object did not connect with the presentation.	Object had some connection with the project.	Object had a clear connection to the presentation.	Object had clearly defined connection with the presentation that was explicitly articulated.

Notes

REFERENCES

Doyle, A. (2019). *IT jobs: Career options, job titles, and descriptions*. The Balance Careers. https://www.thebalancecareers.com/list-of-information-technology-it-job-titles-2061498

King, M. (2018, August 8). *Why we need literature to teach tech*. National Council of Teachers of English. https://ncte.org/blog/2018/08/why-we-need-literature-to-teach-tech/

Office of Educational Technology. (2017). *National education technology plan*. https://tech.ed.gov/netp/

PA Virtual Charter School. (2017, October 4). *How to use your webcam with windows 10 camera app | Tech* [Video]. https://www.youtube.com/watch?v=YFtbPUA1T1o

Shetterly, M. L. (2016). *Hidden figures: The American dream and the untold story of the Black women mathematicians who helped win the space race*. William Morrow.

U.S. Department of Education. (2017). *Reimagining the role of technology in education: 2017 national education technology plan update*. https://tech.ed.gov/files/2017/01/NETP17.pdf

Index

Note: Page numbers followed by *f* and *t* represent figures and tables respectively.

Academic-related disability, 1
Accommodations, in virtual environment, 29
Achievements, 18
Actuaries, 64
Adams, N. M., 63
Adeyemi, T., 24
Adolescents, 35
    culturally and linguistically diverse, 1
    with disabilities, 1, 5–7
    engineering for, 85–86
    literature, 24–25, 24*t*
    mathematics and, 63
    science and literature for, 35
    urban, supporting, 4
Aerospace engineering, 103–106
Aerospace engineers, 86
Aeschlimann, B., 6
African Americans, 3, 18
After Earth (David), 50–52
Agrawal, J., 6
Albert, J. L., 7
Alfeld, C., 25
Algozzine, R. F., 1
Al-Harthy, I., 85
Allen, C., 26
Allen, D. E., 13, 14
Alsbury, T. L., 7
Amazing Stories (series), 35
Anderson, T., 54
Andrews, G., 35
Andriola, D., 6
Animals, desert plants and, 57–58
Application, ATTACK mnemonic, 30, 38, 40–41, 43–44, 45–46, 48–49, 50–51, 53, 54–55, 58, 70. 72, 76, 78, 79, 80–81, 88, 90–91, 93, 94
Archer, A. L., 6, 7, 14
Archer, L., 35
Artiles, A. J., 3
Assaraf, O. B. Z., 35

Assessment, 26–27, 28, 31, 40, 45, 47, 50, 52, 54, 56, 60, 72, 75, 77, 78–79, 80, 81, 89, 92, 93, 95
    formative, 27
    rubric, 91
    summative, 27, 43
Assistive technology, 58–60
Assistive technology innovations, 127–129
Asynchronous activities, 29
ATTACK mnemonic
    application, 30, 38, 40–41, 43–44, 45–46, 48–49, 50–51, 53, 54–55, 58, 70, 72, 76, 78, 79, 80–81, 88, 90–91, 93, 94
    assessment, 31, 40, 43, 45, 47, 50, 52, 54, 56, 60, 72, 75, 77, 78–79, 80, 81, 83, 89, 92, 93, 95
    cognitive reflection, 31, 40, 43, 45, 48, 50, 52, 54, 56, 60, 72, 75, 77, 79, 80, 81, 83, 90, 92, 94, 95–97
    keep, retain, generalize, 31, 40, 43, 45, 48, 50, 52, 54, 56, 60, 72, 75, 77, 79, 80, 90, 92, 94
    teaching strategy, 30, 38–39, 41, 44–45, 46, 49, 51, 53, 55, 58–59, 70, 72–73, 76, 78, 79, 81, 82, 88, 91, 94–95
    trial/try-out, 31, 39, 41–43, 45, 47, 49–50, 51–52, 53–54, 55–56, 59–60, 71–72, 73–75, 76–77, 80, 81, 83, 88–89, 91–92, 93, 95

Back to the Future (Zemeckis and Gale), 107–109
Baeten, M., 13, 14
Barber, B. R., 18
Barnett, S., 16
Barringer, M. D., 25, 27, 28, 35
Barrows, H. S., 13
Basic algebra, and statistics, 70–72
Basilaia, G., 29
Beal, C. R., 63
Bellanca, J. A., 14
Beowulf, 38
Bernhardt, S. A., 13, 14
Berry, R. A., 28
Berry, R. Q., 85
Berry, T. A., 18, 19

Best, A. M., 3
Billingsley, B., 25, 27, 28, 35
Binet, Alfred, 3
Biochemical engineers, 86
Biochemists, 36
Biofuels, 107–109
Biography-driven instruction (BDI), 18
Biophysicists, 36
Blackman, H. P., 2
Blackorby, J., 85
Black Panther (film), 44
Blanchard, M. R., 7
Blanchett, W., 3
Books, literature
    in engineering, 87*t*
    in mathematics, 64*t*
    in science, 37*t*
Born a Crime (Noah), 72–75
Boyd-Batstone, P., 1
The Boy Who Harnessed the Wind (Kamkwamba and
    Mealer), 43–45
Breunlin, R., 2, 3, 4
Brownell, M., 25, 27, 28, 35
Brown-Jeffy, S., 4, 5
Burke, C., 35
Burns, J. B., 14
Burns, M., 1
Business plan, 72–75
Butkevich, E., 3
Butler, S. E., 86
Buysse, V., 2

Calculus, geometry and, 75–77
Camara, W., 1, 6
Cameron, J., 99
Campbell-Whatley, G. D., 3
Card, O. S., 48
Careers
    in engineering, 86
    in mathematics, 64
    in science, 35–37
    technology-related, 111–112
Caste system, 82
Causton-Theoharis, J., 2
Cawley, J., 1
Cells, science on, 40–43
Chemical engineers, 86
Chemists, 36
Cheuk, T., 1
Chinn, P., 3
Christensen, L., 5
The Chronicles of Narnia (Lewis), 79–80
Citations, use of, 114–116
Civil engineers, 86

Clarke, B., 2
Classrooms
    inclusion of diverse students into, 2
    midsized urban school district, 4
    PBL, 13–14
CLD students. *See* Culturally and linguistically diverse
    (CLD) students
Clean water, science and, 43–45
Clemens, N., 4
Cochran, C., 6
Cognitive engagement, 13
Cognitive reflection, 31, 40, 43, 45, 48, 50, 52, 54, 56, 60,
    72, 75, 77, 79, 80, 90, 92, 94, 95–97
Cohen, P. R., 63
Collins, B. A., 1
Collins, S., 24
Comer, J., 3
Common Core State Standards (CCSS), 27–28
Comprehension, reading, 23
Computer hardware engineers, 86
Computer programmers, 111
Computer systems administrators, 111–112
Computer systems analysts, 111
Computer systems engineers, 86
Conservationists, 36
Coogler, R., 44
Cooper, J., 4, 5
Cosier, M., 2
Cotton, K., 16
Coutinho, M. J., 3
COVID-19 pandemic, 35
    and respiratory system, 51
Cox, B. D., 63
Cox, C., 25
Credible sources, identification of, 112–114
Cross-disciplinary learning, 1–2
Cryptographers, 64
Cultural competence, 4–5
Culturally and linguistically diverse (CLD) students, 1, 17,
    85–86
    applicability of explicit instruction to, 15, 15*t*
    biography-driven instructions for, 18
    educational experiences, 17
    and multitiered support systems, 3–4
    and response to intervention, 3–4
Culturally relevant pedagogy (CRP), 4
    concepts of, 17
    outlines foundations for, 5
    positive outcomes of, 17*f*
Culturally responsive problem-based learning (CRPBL),
    13, 14, 18–19
Culturally responsive teaching, 17–18
    student voice using, 18
Culture, 35

Dana, N. F., 14
Darden, Christine, 75
Darling-Hammond, L., 1, 17, 18
Dashner, J., 24, 88
Database administrators, 111
David, P., 50
Dehaene, S., 6
Dehaene-Lambertz, G., 6
Dei, G. J., 17
Delpit, L., 17, 18
Deno, S. L., 3
Desert plants and animals, 57–58
DeWitt, J., 35
Dick, P. K., 103
Digital citizenship, 121–123
Digital footprint, 125–127
Digital media, making use of, 116–118
Digital storytelling, 129–130
Dillon, J., 35
Dimaline, C., 24
Disproportionality, 3
Distance learning, 29–30
Dochy, F., 13, 14
Donham, R. S., 13, 14
Dor-Ziderman, Y., 16
The Dreamkeepers (Ladson-Billings), 17
Drone engineering, 92–94
Drugs, 45–48
Dubois, J., 6

Ecologists, 36
Economists, 64
Education
    goal of, 23–24
    response to virtual learning, 29–30
Educational gaps, 18
Education for All Handicapped Children Act (EHA), 3
Efstathaidis, H., 85
Electrical engineers, 86
Emran, A., 35
Ender's Game (Card), 48–50
Energy, 54–56
Engineering
    for adolescents, 85–86
    aerospace, 103–106
    biofuels, 107–109
    careers in, 86
    cyborgs, 99–101
    drone, 92–94
    satellite, 97–99
    software, 101–103
    and space travel, 106–107
    tips for teachers, 86–87
    trains, 94–97

Engineering construction and building, 90–92
Engineering robotics, 88–90
English as a second language (ESL), 1
English learners (ELs), 1, 4
    brain function, 6
Environmental engineers, 86
Erduran, S., 35
Eshach, H., 16
Explicit instruction, 14–15, 15t

Feistritzer, C. E., 4
Fernstrom, P., 2
Financial planners, 64
Flatland, 82
Flegg, R. B., 35
Flight engineers, 86
Fogarty, R. J., 14
Foley, T. E., 1
Ford, M. R., 85
Forensic scientists, 36
Formative assessments, 27
Forness, S. R., 1
Foster, A. D., 58
Francis, D. J., 15
Frankenstein, 35, 38
Freire, P., 14, 17
Fuchs, D., 1, 2
Fuchs, L. S., 2, 5
Full inclusion programming, 2
Furner, J. M., 25

Gagnon, J. C., 18
Gale, B., 107
Garza, R., 4
Gattaca (Niccol), 53–56
Genetics, 53–54
Geometry, 80–83
    and calculus, 75–77
Geoscientists, 36
Gey, G., 40
Gherasoiu, I., 85
Gijbels, D., 13
Girls, scientific careers for, 35
Giroux, D., 2, 3, 4
The Giver (Lowry), 78–79
Global Scholar, 3
Goeke, J., 6, 7, 14
Gog, T., 13
Goodard, H. H., 3
Google Docs, 29
Gouleta, E., 3
Grade point averages (GPAs), 85
Gravity, 48–50

Greene, K., 15
Gronneberg, J., 25

Haager, D., 5
Hammond, L., 14
Harry, B., 3
Harry Potter (Rowling), 24, 25, 90–92
Heick, T., 16
Henley, M., 1
Henriksen, E. K., 35
Heron, T. E., 3
Herrera, S., 18
Herrmann, Z., 29
Hertz-Pannier, L., 6
Herzog, W., 6
Heward, W. L., 3
Hidden Figures (Shetterly), 24, 75–77
Hieb, J. L., 26
High-leverage practices (HLPs), 28
Hilt-Panahon, A., 4
Hitchcock, C., 1
Holes (Sachar), 7, 24, 57–58
Honey, M., 24
Hong, S., 3
Houchins, D. E., 18
Huerta, M., 3
Hughes, C., 6, 7, 14
The Hunger Games (Collins), 24, 70–72
Hunt, J., 2, 3, 4
Hurd, G. A., 99
Hydro/hydraulic/water engineers, 86
Hydrologists, 37

The Immortal Life of Henrietta Lacks (Skloot), 40–43
Inclusive procedures, 2
Individuals with Disabilities Education Act (IDEA), 2
Industrial engineers, 86
Information security analysts, 111
Inquiry-based learning, 15–16
    student-directed, 16–17
Intelligence test, 3
Interdisciplinary classroom
    STEM teaching in, 24–27
    teaching strategies and activities, 25–26
Investment analysts, 64
Isaacson, R. M., 85

Jackson, D., 25, 27, 28, 35
Jackson, M., 75
Jackson, Y., 18
Jennings, S., 86
Jeon, H.-J., 3
Johnson, D. W., 13
Johnson, K., 75

Johnson, R. D., 23
Johnson, R. T., 13
Johnston, S., 25
Jones, S., 13
Jumanji (Allsburg), 101–103

Kagan, S., 25
Kamkwamba, W., 26, 43
Kang, E., 38, 87
Karier, C. J., 3
Kartheiser, G., 6
Kavale, K. A., 1
Kennedy, M., 25, 27, 28, 35
Keollner, K., 63
King, J. E., 18
Kirby, M., 2
Klingner, J., 3
Kloo, A., 4
Kozol, J., 18
Kvavadze, D., 29
Kwon, K., 3
Kyndt, E., 13, 14

Lackaye, T. D., 2
Lacks, H., 40
Ladson-Billings, G., 4, 5, 17
LaForest, R. K., 85
Langdon, C., 6
Latinxs, 3, 18
Learning
    inquiry-based, 15–17
    problem-based, 29–30
    virtual, 29–30
Lee, H., 24, 65
Lee, O., 1
Leone, P. E., 18
Lesaux, N. K., 15
Lesson planning, 30–31, 31*f*
    engineering, 87–109
    mathematics, 64–83
    science, 38–60
    technology, 112–131
Levithan, D., 24
Lewis, C. S., 79
Lewis, T., 25, 27, 28, 35
Linnenbrink, E. A., 13
Literature
    adolescent, 24–25, 24*t*
    benefits of, 5–7
    engineering and. *See* Engineering
    mathematics and. *See* Mathematics
    science and. *See* Science
    and STEM, benefits of connecting, 23–24
    technology and. *See* Technology

Liu, S. Y., 35
Losen, D., 18
Lott, K., 24
Lower order questions, 16, 16f
Lowry, L., 78
Loyens, S., 13
Lynch, S. J., 26, 85

Maheady, L., 25, 27, 28, 35
Mahzoon-Hagheghi, M., 23
Makarova, E., 6
Maker, C. J., 85, 86
Mantzicopoulos, P., 85
Margalit, M., 2
Marine engineers, 86
Masnick, A. M., 63
Mass, 48–50
Mathematicians, 64
Mathematics
   and adolescents, 63
   business plan, 72–75
   careers in, 64
   geometry and calculus, 75–77
   lesson planning, 64–83
   literature books in, 64t
   mysticism of the pyramids, 78–79
   statistics and probability, 65–70
   teacher application tips, 63
   weather, 79–80
The Matrix (Wachowskis and Wachowskis), 54–56
Mattern, K., 85
Mayer, D., 24
The Maze Runner (Dashner), 24, 88–90
McClain, O. L., 85
McDonough, I. M., 63
McIntyre, J. G., 86
McLeskey, J., 25, 27, 28, 35
Mealer, B., 26, 43
Means, B., 26
Mechanical engineers, 86
Mechatronics engineers, 86
Medicine, 45–48
Men in Black (Perry), 92–94
Mesci, G., 37
Metacognition, 31
Meyer, A., 1
Meyer, M., 24
Meyer, S., 24, 38
Mikkers, J., 13
Milner H. R., 17
Minero, E., 29
Mining and geological engineer, 86
Minority Report (Dick), 103–106
Mkhize, D. R., 63

Mock, D., 1
Models of service, 2
Morales, E. E., 85
Morgan, P. L., 1
Morin, L. L., 6
Morpheme, 26, 26f
Morrison, D., 2, 3, 4
Motivation, 6
Moughamian, A. C., 15
Multitiered support systems (MTSS), 3–4, 15
Murry, K., 18
Mysticism of the pyramids, 78–79

Nanotechnology engineers, 86
National Association of School Psychologists, 3
National Center for Education Statistics, 1
National Education Technology Plan (NETP), 27
Neumann, D. L., 35
Niccol, A., 53
Noah, T., 72
Nuclear engineers, 86

Odom, S., 2
Online learning, 29–30
Operations research analysts, 64
Ortiz, A., 2
Osman, C. J., 63
Oswald, D. P., 3

Parmar, R., 1
Patrick, H., 85
PBL. See Problem-based learning (PBL)
Pearson, D., 25
Pearson, G., 24
Pease, R., 85, 86
Peer-assisted learning strategies (PALS), 25
Percy Jackson and the Olympians series (Riordan), 24
Percy Jackson and the Lightning Thief (Riordan), 45–48
Perry, S., 92
Pete, B. M., 14
Peters, V., 26
Petitto, L. A., 6
Pleasants, B. A. S., 37
Prasse, D., 2, 3, 4
Price, Y., 24
Probability, statistics and, 65–70
Problem-based learning (PBL), 7, 29–30
   best practices, 14
   characteristics of, 13
   culturally responsive. See Culturally responsive
      problem-based learning (CRPBL)
   goal of, 13
   innovations in instruction for, 14–17
   instructions, 13, 14

Problem-based learning (PBL) *(continued)*
  skills need to implement, 14

Quenemoen, R., 1, 6
Questioning
  lower order, 16, 16*f*
  for teachers, 24
  to/by students, 15–16, 16*f*
  true-or-false, 29
  types of, 18
Quinn, H., 1

Raborn, D. T., 63
Radunzel, J., 85
Ralston, P. S., 26
Ramirez, G., 63
Ramsey, R. S., 1
Reading comprehension, 23
Reading disability, 1
Reilly, D., 35
Reschly, D., 3
Respiratory system, 50–52
Response to intervention (RTI), 3–4
Reynolds, J. A., 24
Ring, E., 2
Riordan, R., 24, 45
Rivera, M. O., 15
Rivoli, G., 26
Roache, M., 3
Robotics engineers, 86
Rodenberry, G., 97
Rodriguez, D., 3
Rodriguez, J., 25, 27, 28, 35
Rogers, C., 5
Roghaar, D., 24
Rogow, F., 30
Romance, N. R., 85
Root-Bernstein, R., 5
Rose, D., 1
Rowling, J. K., 24
Roy, S., 1
Rubric, 40
  assessment, 77, 91
  student academic, 90
  for summative assignment, 43
Ryder, J., 35

Sachar, L., 7, 24, 57
Sáenz, L. M., 5
Salmon, S., 1
Samarapungavan, A., 85
Samson, J. F., 1
Santos, M., 1
Satellite engineering, 97–99
SAT scores, 85

Scheeler, M. C., 25, 27, 28, 35
Scheibe, S., 30
School inequities, 18
Schwartz, R. S., 37
Schwartzbach-Kang, A., 38, 87
Schweingruber, H., 24
Science
  for adolescents, 35
  assistive technology, 58–60
  career choices, 35–37
  on cells, 40–43
  and clean water, 43–45
  desert plants and animals, 57–58
  energy, 54–56
  genetics, 53–54
  and literature, 25, 35–60
  and medicine/drugs, 45–48
  respiratory system, 50–52
  skills, books and, 37*t*
  teacher application tips, 37–38, 37*t*
  vocabulary, 38–40
Sedita, J., 26
Segers, M., 13
Shabatura, J., 13
Shaprio, E., 4
Shattuck, P., 85
Sheffield, S., 1
Sheppard, S. D., 13
Shetterly, M. L., 24, 75
Shi, Y., 85
Shifrer, D., 2
Shippen, M. E., 18
Shore, J., 3
Shumn, J. S., 2
Sickle-cell anemia, 53
Singh, N. N., 3
Skloot, R., 40
Skordoulis, K., 35
Smith, K. A., 13
Snowpiercer (Lob), 94–97
Social and emotional learning (SEL), 30
Software developers, 111
Software engineering, 101–103
Software engineers, 86
Sohn, L. N., 23
Soukakou, E., 2
Space travel, engineering and, 106–107
Special education law, 3
Spektor-Levy, O., 35
Spillane, N. K., 85
Stack-Oden, M., 85
Stanfa, K., 4
Star Trek (series), 35
Star Trek—The Motion Picture: A Novel (Rodenberry), 97–99
Star Wars: Splinter of the Mind's Eye (Foster), 58–60

Statisticians, 64
Statistics
    basic algebra and, 70–72
    and probability, 65–70
STEM
    benefits of, 5–7
    careers, 7
    content, adolescent literature pertaining to, 24–25, 24t
    innovations in instruction for PBL in, 14–17
    literature and, benefits of connecting, 23–24
    teaching in interdisciplinary classroom, 24–27
Stone, A., 6
Stone, J. R. I., 25
Stowe, N., 3
Strobel, J., 7, 13
Struyven, K., 13, 14
Student-directed inquiry-based learning, 16–17
Students with disabilities, 1, 18
    and academic-related disability, 1
    applicability of explicit instruction to, 15, 15t
    CLD, 2–4
    as English learners, 1
    high-leverage practices and strategies for, 28
    STEM texts characteristics for, 5
    strategies for, 2
    in urban environment, 4–5
Students with special needs, 2
Summative assessments, 27, 43
Swackhamer, L., 63
Synchronous activities, 29

Tal, O. P., 35
Tampakis, K., 35
Tankersley, K., 24
Taplin, A., 29
Tate, W., 17
Tay-Sachs disease, 53
T-chart, 42
Teachers
    application tips, 37–38, 37t, 63, 86–87, 112
    and asynchronous activities, 29
    culturally relevant, 17
    in inclusive settings, 2
    proficiency, tips for, 16
    questions for, 24
    recommendations for, 19
    and synchronous activities, 29
    and teaching strategies and activities, 25–26
Teaching
    activities, 25–26
    strategies, 25–26, 30, 38–39, 41, 44–45, 46, 49, 51, 53, 55, 58–59, 70, 72–73, 76, 78, 79, 81, 82, 88, 91, 93, 94–95
Technology, 26, 27, 31
    assistive, 58–60

assistive technology innovations, 127–129
    credible sources, identification of, 112–114
    digital citizenship, 121–123
    digital footprint, 125–127
    digital storytelling, 129–130
    lesson planning, 112–131
    and literature, 111–131
    making use of digital media, 116–118
    and online learning, 29
    teacher application tips, 112
    3D print, 130–131
    to update writing products, 119–121
    use of citations and presentations, 114–116
    webcams, 123–125
Terman, M., 3
The Terminator (Cameron and Hurd), 35, 99–101
Theoharis, G., 2
Thier, K., 2, 3, 4
3D print, 130–131
Thunder, K., 85
Thurlow, M., 5
Time, 30, 41
To Kill a Mockingbird (Lee), 24, 65–70
Trains, engineering, 94–97
Travers, J., 2
Tremblay, P., 2
Trial/try-out, ATTACK mnemonic, 31, 39, 41–43, 45, 47, 49–50, 51–52, 53–54, 55–56, 59–60, 71–72, 73–75, 76–77, 78, 80, 81, 83, 88–89, 91–92, 93, 95
Tsai, C. C., 35
Turnbull, R., 3
Twilight (Meyer), 24, 38–40
Twilight Zone (series), 35

United Nations Educational, Scientific, and Cultural Organization (UNESCO), 29
Universal Design for Learning (UDL), 25
Urban environment, students with disabilities in, 4–5

Valdés, G., 1
Valenti, S. S., 63
Van Barneveld, A., 13
Van den Bossche, P., 13
Van Loan, C., 18
Vaughn, Dorothy, 75
Vaughn, S., 2, 5
Virtual learning, 29–30
Vitale, M. R., 85
Vlahakis, G. N., 35
Vocabulary, 26, 30, 41, 42, 44, 46, 49, 51, 52, 53, 55, 59, 70, 71, 73, 76, 78, 80, 81, 88, 91, 93, 95
    science, 38–40

WADE mnemonic, 70
Wallace, F. H., 63

Wallin, M., 24
Wang, H., 26
Was, C. A., 85
Weather mathematics, 79–80
Webcams, 123–125
Web developers, 111
Wei, X., 26, 85
Weight, 48–50
Weir, A., 106
Welner, K., 18
Westrick, P., 85
White, D., 85
Williams, B., 7
Winn, J., 25, 27, 28, 35
Witzel, B., 2
Wolkenhauer, R., 14
Wolpert-Gawron, H., 15
"Words Alive" map, 26, 27f

Wright, G. A., 23
Wu, I. C., 85, 86

Yates, J., 2
Yebra, R., 23
Yefroimsky, Y., 16
Yoon, S. Y., 7
Young, C. L., 1
Young, V., 26
Yu, J. W., 85

Zemeckis, R., 107
Zeyer, A., 35
Ziegler, D., 25, 27, 28
Zigmond, N., 4
Zoboi, I., 24
Zunon, E., 26

About the Authors

Dr. Gloria Whatley is a full professor in the Department of Special Education and Child Development at the University of North Carolina at Charlotte. She serves as a teacher of graduate students there and has served as graduate coordinator. She received her BA from Dillard University, her MA at the University of Alabama at Birmingham, and her doctorate of education degree at the University of Alabama in Tuscaloosa. She has worked as a classroom and learning disability teacher. In higher education, she has served as a director and has chaired a special education program. She chaired and started the special education program at Indiana University-Purdue University Fort Wayne. She has served as a program specialist in the central office for Birmingham City Schools' special education programs in Alabama. She has also been on the Council for Exceptional Children's national and international boards. She has delivered numerous national and international presentations, workshops, and strands. Her specialty is infusing diversity into higher education and K–12 curriculum, and she also offers solutions for behavior problems, response to intervention, and social skills training in public schools. Dr. Whatley has written several articles related to inclusion and published two books on behavior. She has cumulatively written four books, including the textbook *Leadership Practices for Special and General Educators* (2013). She also coauthored *A School Leader's Guide to Implementing the Common Core: Inclusive Practices for All Students* (2016). Her research focuses on diversity, social skills and behavior, and administration in special education. Recently she received the Diversity Award for the UNC Charlotte Cato College of Education and she served as a fellow providing diversity training for the 17 schools in the University of North Carolina System.

Dr. Diane Rodríguez is a full professor at the Fordham University Graduate School of Education. Her research interconnects special education, bilingual education, and the academic development of culturally and linguistically diverse students. Dr. Rodríguez has been an invited speaker at national and international conferences on special education and bilingual education. The Spanish-language television network Univision selected Dr. Rodriguez as an example of "Orgullo de Nuestra Comunidad," a designation that honors individuals who give back to the community. She was recognized for her work with individuals with disabilities. She has written extensively in journals and is the co-author of *The Bilingual Advantage* (2014), published by Teachers College Press. Dr. Rodriguez is the founder of Every Girl Is Important, a not-for-profit organization that works to dismantle barriers and help girls attain secondary education achievement that can change their lives, villages, countries, and the world.

Dr. Jugnu Agrawal is a program manager for special education curriculum with Fairfax County Public schools and an adjunct professor in the College of Education and Human Development at George Mason University in Virginia. She received her BSc and MSc degrees in child development from Lady Irwin College in India and a second master's degree in special education and a doctor of Education at George Mason University. She has taught in K–12 settings for over 20 years, working with students with intellectual disabilities, autism, and learning disabilities in India and the United States. She has also worked as a consultant in India for two school systems, assessing students and training teachers on working with students with learning disabilities. In her current job, she provides training and support related to curriculum and behavior to special education teachers. As an adjunct professor, she has taught several graduate-level courses on strategies for teaching students with learning disabilities, positive behavior supports, consultation and collaboration, and research in special education. Dr. Agrawal is the past president of the Division of Special Education Services of the Council for Exceptional

Children. She has also served as the chair of the diversity committee for the Council for Learning Disabilities. Her research interests include international special education, mathematics interventions, and the use of technology to support students with disabilities. She is passionate about closing the research-to-practice gap by supporting teachers in the use of evidence-based interventions in K–12 settings. She has presented at several national and international conferences and workshops and has published several articles on learning disabilities.